The Book of Looms

The Book of Looms

A history of the handloom from ancient times to the present

Eric Broudy

BRANDEIS UNIVERSITY PRESS
Waltham, Massachusetts

FOR DAISY — *AT LAST!*

Brandeis University Press
© 1979 Eric Broudy
All rights reserved
Manufactured in the United States of America

First Brandeis University Press edition 2021
Previously published by University Press of New England
in 1993
Originally published by Van Nostrand Reinhold in 1979

Brandeis paperback edition ISBN: 978-1-68458-082-8
Brandeis ebook edition ISBN: 978-1-68458-083-5

For permission to reproduce any of the material in this
book, contact Brandeis University Press, 415 South
Street, Waltham MA 02453
or visit brandeisuniversitypress.com

Library of Congress Cataloging-in-Publication Data
Names: Broudy, Eric, author.

Title: The book of looms : a history of the handloom from
    ancient times to the present / Eric Broudy.

Description: First Brandeis university Press edition. |
    Waltham, Massachusetts : Brandeis University Press,
    2021. | Includes bibliographical references and index.|
    Summary: "Broudy shows how virtually every culture,
    no matter how primitive, has woven on handlooms.
    He highlights the incredible technical achievement of
    primitive cultures that created magnificent textiles with
    the crudest of tools and demonstrates that modern
    technology has done nothing to surpass their skill or
    inventiveness" — Provided by publisher.

Identifiers: LCCN 2021018112 (print) | LCCN
    2021018113 (ebook) | ISBN 9781684580828
    (paperback) | ISBN 9781684580835 (ebook)

Subjects: LCSH: Looms.

Classification: LCC TS1493.B69 2021 (print) |
    LCC TS1493 (ebook) | DDC 677/.02854—dc23

LC record available at https://lccn.loc.gov/2021018112

LC ebook record available at https://lccn.loc.gov/
    2021018113

5 4 3 2

ACKNOWLEDGMENTS

There aren't many, but the few people who have made their presence felt in this book deserve mention here.

Mary Elizabeth King, Keeper of Collections, the University Museum, Philadelphia, acted as formal consultant. She read and annotated every chapter, and her suggestions and corrections saved me much time and countless errors.

Katherine R. Koob, Assistant Curator of Textiles, Merrimack Valley Textile Museum, Andover, Mass., read and criticized Chapter 7, "The Drawloom," and gave freely of her knowledge and understanding of how the looms discussed in that chapter work.

Jerome Jacobson, Assistant Professor of Anthropology, The City College, City University of New York, read and criticized Chapter 1, "Origins," and among other things alerted me to the Seuss calibration chart, which compensates for the inaccuracies of carbon-14 dating beyond 1000 B.C. He tried valiantly to check the many dates to assure consistency in dating, but I take full responsibility for any errors.

Howard K. Battles, a close friend and superb editor, read and criticized the entire manuscript for clarity and style. His thirty-five pages of notes would have devastated me had I not received them a chapter at a time. His comments were invariably on the mark, forcing me to attend to loose writing, vagueness, and inadequate explanations of technical detail.

At various times the following people provided direction, advice, criticism, and support far in excess of what they probably thought, and for their help I am equally grateful: Milton Sonday, Curator, Cooper-Hewitt Museum, New York; Dr. Junius B. Bird, Curator Emeritus, American Museum of Natural History, New York; Rita Adrosko, Curator, Division of Textiles, Smithsonian Institution, Washington, D.C.; Ruth Katz, author of *Card Weaving*.

Finally, I would like to express my gratitude to the New York Public Library for making the resources of the Frederick Lewis Allen Room available to me during my many long months of research.

Contents

Introduction

The Book of Looms had its origin in simple curiosity. In assembling various looms for my wife I found myself wondering about the lineage of what appeared to be a most simple yet elegant tool. I consulted the few weaving books she had acquired, but in them, as in most others I later investigated, the loom was treated only incidentally—usually in a "how-to-weave" context, rarely historically. I learned that, in general, writers on weaving concerned themselves primarily with textiles, drafts, yarns, colors, and patterns and discussed looms only as much as was necessary to describe how to tie on a particular pattern. Any discussion of looms themselves was relegated to the murky journals of archaeologists and anthropologists who viewed the loom as evidence of a certain state of cultural development or pattern of migration in ancient or primitive societies. The archaeological material is highly specialized and fragmented, and only a few have tried to piece it together and even then only for a limited geographic area.

M. D. C. Crawford noted that weaving is "the most ancient of the great arts," appearing at the dawn of history, virtually inseparable from true culture. Crawford wrote, "From the rough fish weirs to the most elaborate baskets, from the coarser fabrics of flax to the gossamer webs of cotton and silk, it has sustained and beautified (man's) life from the night of history to the latest passing hour; it is the veritable nurse of civilization." Crawford was not wrong. The principles, the tools, even the language of weaving have acquired by their fundamental importance symbolic and metaphoric value in our lives. It is said that in China the warp, firmly fastened to the loom, symbolizes the immutable forces of the world, while the weft, shifting back and forth, symbolizes the transient affairs of man. In India the warp represents eternal existence, and the weft symbolizes the stages of an individual's life. Our word "heirloom" originated with the family loom that was passed down from generation to generation, the word "spinster" derived from the custom that unmarried women spun a certain quantity of yarn before they were married for making the household linens. The customs, practices, and language surrounding the loom and its products over the centuries have thoroughly woven themselves into the very warp and woof of our culture.

Our word "loom" derives from the Old English geloma, which meant simply "tool" or "utensil." The loom, perhaps next to the stone ax and spear, was the tool in ancient times. Its history has been largely neglected in favor of the textiles woven on it—partly because textiles have survived in greater quantities than looms and partly because the use of a tool can tell us more about a culture than can the tool itself. Nonetheless, the loom has an impressive history that must have been preceded by a prehistory of even greater duration than the period since the earliest textile discoveries.

The Book of Looms carries the story of the handloom up to the present, but I have omitted detailed discussion of developments during the Industrial Revolution. This period during the mid- to late-nineteenth century introduced a number of mechanical "improvements" that took the loom out of the hands of the handweaver and put it in a factory where a relatively unskilled technician could monitor the production of cloth. While I have noted the major advances in loom technology during the industrial age, mechanical looms do not figure in my discussion of the handloom except insofar as they affected the progress of handweaving. Unlike the neglected handloom, mechanical looms have been amply described elsewhere.

The history of the evolution of the loom is a history of minor innovations, mostly designed to increase the speed of fabric production. The entire weaving process can be simplified into three basic operations: holding the warp under tension, opening and changing the shed, and inserting and beating up the weft. All the improvements and changes in loom design and construction are concerned with one or more of these problems. Once weaving entered the commercial arena, doing it better usually meant doing it faster. Today, pattern cards for jacquard weaving can be cut by computer and woven at the rate of 200 picks per minute. On other modern looms water jets can propel weft yarn through a shed at the rate of *1,000 picks per minute!* The handloom, which began as a mechanism to furnish necessities, has survived, at least in western societies, as a specialized tool of the handcraftsman who furnishes art or luxury fabrics.

This history makes no pretense at being exhaustive. Not every loom or loom type from every area has been discussed. In the interest of clarity, I have concentrated on the major types of looms and influences on loom development in areas of the world that offer sufficient evidence to link segments of the loom's fragmentary history together. I have arbitrarily omitted carpet and rug looms from the discussion because of their specialized techniques of knotting, even though their loom frames closely resemble those used for cloth weaving. I have tried to contain my discussion as much as possible to looms, referring to textiles, spinning, costume design, color, and so on only as far as was needed to illuminate loom design and construction. There are obvious dangers in this approach, and some readers may object to certain generalizations that result from narrowing my focus to the tool itself. I expect that the reader will want to know more about textile history, trade routes, and costumes of various eras, but this information is available elsewhere and would only smother the story of the loom itself if presented here.

It pays to remember that it is the historian, not history itself, that organizes the past for the benefit of the present reader. The history of the handloom is nowhere as neat and sequential as I have presented it here. Many elements of its history are as yet unknown, and many of the distinctions I make may collapse before the study of future or more experienced historians. I regard *The Book of Looms* as an initial endeavor—a book that is comprehensive enough for the lay reader, specialized enough for the archaeologist or textile historian, and I would hope, accurate and readable enough for both.

1. Origins

Spider Woman instructed the Navajo women how to weave on a loom that Spider Man told them how to make. The crosspoles were made of sky and earth cords, the warp sticks of sun rays, the heddles of rock crystal and sheet lightning. The batten was a sun halo; white shell made the comb.—Navajo legend

THE ANTIQUITY OF WEAVING

No one knows nor is ever likely to know how weaving began, but the idea of weaving clearly preceded the loom by many thousands of years. The legends of most cultures, with the exception of the Chinese, place the invention of weaving at the beginning of their own history. That it is often attributed to the gods testifies to its importance to these ancient cultures. The Peruvians credit Mama Ocllo, the wife of Manco Capac, their first sovereign. The Assyrians honor Queen Semiramis, the Egyptians the goddess Isis—usually pictured with shuttle in hand. (According to Egyptian mythology, flax, the fiber associated with the finest Egyptian weaving, was the first thing that the gods created for themselves before appearing on earth.) The Mohammedans believe that weaving originated with the son of Japheth, the third son of Noah, who, according to some traditions, was the ancestor of the Indo-European race. In Greece the honor belongs to Athena (in ·Rome Minerva), the goddess of the arts, who is often represented holding a distaff.

The real origins of weaving are obscurely nestled in what ethnologists call the "dawn of prehistory." The image is an apt one because it suggests that weaving has been with us for about as long as we would care to consider primitive man our ancestor. In that early dawn of modern man, probably in the Upper Paleolithic Age when much of North America and Eurasia was still covered with glacial ice and the cave bear and mammoth still wandered the earth, the first seeds of weaving began to germinate. Some of the technology that the weaver would later put to use already existed. Upper Paleolithic man dressed himself in skins sewn together with bone needles. For thread he used animal sinews and intestines and a variety of reeds, rushes, and bast fibers that he had previously learned to spin.

Some authorities believe that spinning preceded weaving by a full cultural cycle. The earliest spun fibers probably were used to carry or pull things or to fasten stones to sticks to form axes and other weapons. This early cordage was later adapted to fishnets, snares, slings, bowstrings, and other simple uses. The imagination suggests various ways in which spinning may have originated—perhaps from observing certain vines that twined about each other, or by watching the action of a stone ax head as it slipped from the fingers and twirled the fibers to which it had already been knotted, or by idle experimentation. One authority has even asserted that man has "an innate desire to play with fibers and string."

Other authorities claim that spinning did not necessarily precede the invention of weaving. One noted anthropologist, Alfred Bühler, explained: "In looms and woven fabrics of the most primitive type warp and weft consist of knotted threads, and in many places such material is used exclusively. It would appear more probable that the knots, which formed a handicap for weaving, led to efforts to find more suitable threads and thus to the invention of the technique of spinning." How spinning actually began doesn't concern us here, but we may assume that some knowledge of fiber preceded weaving just as the knowledge of weaving preceded the loom.

If we can only speculate on the origins of spinning, so too we can only suppose how man first got the idea to weave. Perhaps it was from certain birds that weave nests (fig. 1-1) or from watching how the wind interlaced the leaves of the date palm. While the legends of most cultures attribute the origins of weaving to the gods, many of the

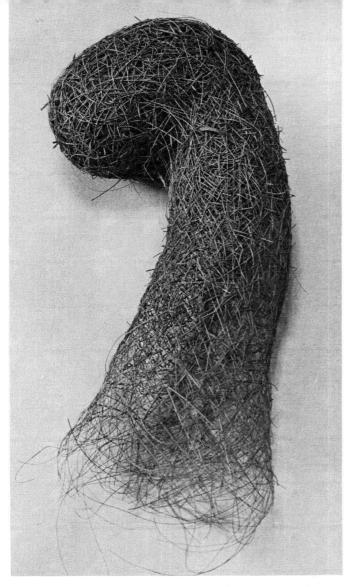

1-1: Weaver-bird nest from West Africa. Collection of the author. Photo by Barbara Wrubel.

same legends support the notion that weaving was linked with what man found in nature. In *The Art of the North-East Frontier of India* Verrier Elwin relates the Indian legend of the Kaman Mishmis:

> Originally people did not wear clothes, for they did not know how to weave. The first weaver was a girl named Hambrumai, who was taught the art by the god Matai. She sat by the river and watched the waves and ripples on its surface and imitated them in her designs. She lay in the forest looking up at the patterns woven by the branches of trees, the leaves of the bamboo; she saw ferns and plants and flowers, and from these things learned other designs . . . .

Perhaps man acquired the notion of weaving when he saw how rushes strewn on the dirt floor of a cave tended to work themselves together from the trodding of feet, or perhaps he was inspired, as the following African legend tells us, by the spider:

> Once there was a man who was a great hunter. He fell sick, and as he lay out of doors he saw a big spider making a net on a bush, and he watched him. By and by he saw how the spider caught his prey. After a time he tried to make a net like the spider's out of bush rope [the long, twining plants that grow in the bush]. He did it, and put his net in the forest and caught bush deer and porcupines, and he became a greater hunter than ever.
>
> One day the spider made a fine cloth, and the hunter's wife admired it and said, "This cloth is better than our cloth [bark-cloth]; make me some like it." And the man tried to, but he could not get a good shape into it, so he went to the spider again, and took him an offering and said, "O my lord, teach me more things." And he sat and watched the spider for many days. By and by he saw that the spider made his net *on sticks.* So he went and got new bush-ropes and fixed it on to the bush near the spider, and made a new net, and his wife was much pleased.
>
> By and by the man saw that he did not want all the sticks of a bush to make his net on, only some of them, and so he took these home and put them up in his house and made his nets there. After a time his wife said, "Why do you make the stuff for me with bush-rope? Why do you not make it with something finer?" Again, he went into the bush to see the spider, and made an offering to him, saying, "O my lord, teach me more things." And he sat and watched the spider and saw how the thread came out of his body, so he said in despair, "O my lord, you are greater than I am; I cannot do this thing."
>
> And as he went home, thinking, he saw there were different kinds of bush-rope, and there was grass which was thinner still. So he took the grass and made a net with it, and he made more nets, and every net was better than the last. His wife was really pleased now, and said, "This is good cloth." The man lived to be very old, and was a great hunter and a great chief.

If a spider inspired the African, then in China we must look to a caterpillar, the *Bombyx mori,* commonly known as the silkworm. According to one version of this legend the prince Hoang-ti wanted his wife, Si-ling-chi, to contribute to the happiness of his people. He gave her the responsibility of studying the silkworm to see if there were a way to make its thread usable. She collected some of the silkworms, fed them herself and learned how to raise them. It is said that by accidentally dropping a cocoon into boiling water, she learned the secret of ungluing the filaments from one another. Besides discovering how to reel the silk from the cocoon, she is credited with inventing the loom (c. 2640 B.C.) and has since been deified as the Goddess of Silkworms.

The spider and caterpillar, the ripples in water, and branches interlaced in the trees—the legends tend to support the most frequently quoted scenarios, that nature itself planted the first seeds of weaving. One might suppose that the first fish weirs derived from the natural tangling of branches in the narrow sluices of rivers; the first wattled windbreaks may have been suggested by the dense growth of the trees themselves. Sometime in this early predawn the mist did lift, and man found himself in possession of a revolutionary idea.

The first weaver may have been the man who first thought to improve on nature's fish weirs by interlacing more branches in a tree that had fallen into and partially blocked a river. Having grasped that idea, he may then have endeavored to provide some protection for himself while he fished by planting some poles in the ground and twining more supple branches between them.

The temptation in discussing an evolutionary process, once you have established a plausible beginning, is to seek links that lead in direct lines—a spawning b and c, b and c spawning d, e, and f, g, respectively—like a family genealogy down to the present. But this temptation must be resisted for several reasons. A beginning that seems plausible through historical hindsight in fact may not relate at all to how the process originated. Accident must be at least as prevalent in the history of ideas as in the evolution of a species, particularly with regard to an idea as fundamental as weaving. The rudimentary beginnings of weaving undoubtedly originated independently in various places and in various ways.

Elizabeth Siewertsz van Reesema cautioned against accepting too readily the prevalent view that weaving was taken over from nature. In her article "Contributions to the Early History of Textile Tehnics" she stated that the interlaced construction of wattled windbreaks does not imply a relation to the same pattern later found in cloth, for the time separating the two products is far too long. A more reasonable relationship, she claimed, exists between the nature of the material at hand and what is done with it at the time. Given the stiff branches of the windbreak, the interlaced method is not surprising, whereas finer materials could be worked in a number of techniques, such as knotting, looping, and plaiting. In other words, the properties of different fibers require different manipulation, and this relationship is more compelling than that of parallel techniques in different materials. Nonetheless, it is still conceivable that, when finer fibers became available, a wattled windbreak at the edge of a stream may have suggested a method of working those fibers.

BASKETRY AND MAT MAKING

Paying lip service to the legendary origins of weaving, the archaeologist and ethnologist return us to the Mesolithic Age to find what they believe are more immediate ancestors of textile weaving: basketry and mat making. At this time environmental changes began to occur that would profoundly alter the development of cultures. As the glaciers retreated, taking with them the grasses that had adapted to the glacial climate, the reindeer, a staple of primitive man's diet, followed. At the same time the sea level, on the rise from melting glaciers, inundated coastal shelves and estuaries to provide ideal habitats for shellfish, shallow-water fish, and waterfowl. While some hunters followed the venison, other peoples in areas as diverse as the Near East, North America, and Africa began to settle where newer foods were now becoming more abundant —alongside waterways and near migratory bird routes.

The population pressure of growing communities forced each to make do with less land and to exploit what grew wild, swam, or flew in its own backyard. When the population outgrew the food resources, some groups were forced to move into new areas. Perhaps these settlers carried with them some wild grains from the old communities and replanted them where they resettled. In such a way, perhaps only incidentally, was agriculture born. New communities grew up in the area east of the Mediterranean from Greece into Southwest Asia and perhaps in Southeast Asia as well. Community development introduced the "civilized" notions of secure territory and organized trade. The change of diet and a more or less secure dwelling brought with them a change of lifestyle and household duties. The dawn of the Neolithic Age introduced the modern concept of homemaking.

The wattled shelter or windbreak no doubt inspired the wattled houses, later plastered over with mud from the riverbanks. This in turn—and here again we can only speculate—must have led to Neolithic man's first luxuries—woven mats and baskets. The materials for these amenities varied with the place, but they included skins, roots, rushes, palm fronds, and a multitude of grasses, sedges, reeds, and straw—in short, whatever was locally available.

The fibers for the earliest baskets generally were not spun, except for handles and bases, which occasionally were hand-twisted for extra strength. R. J. Forbes in *Studies in Ancient Technology* identifies six types of basketry in ancient times (fig. 1-2). While the history of basketry is beyond the scope of this book, I mention these six types because there is a reasonable certainty that plaiting baskets was a preliminary step to weaving cloth. The consensus is that many weaving patterns are derived from these techniques. For example, baskets exist in tapestry, gauze, twill, and embroidery—patterns that also appear in cloth.

1. *Coiled* basketry, probably the oldest technique, consists of bunches of fibers coiled spirally ring upon ring until the desired height is reached. As the coils are laid down, they are fastened by sewing the top coil to the one underneath with fibers of the same material.

1

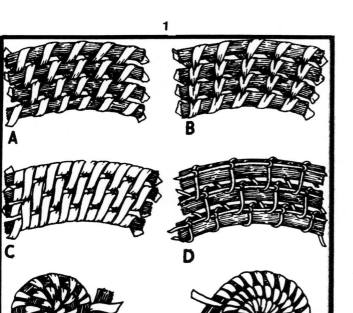

2 3

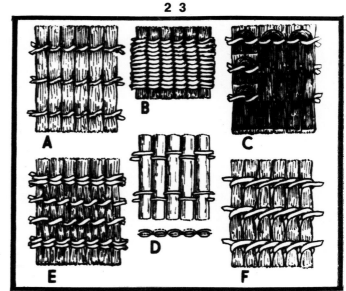

4

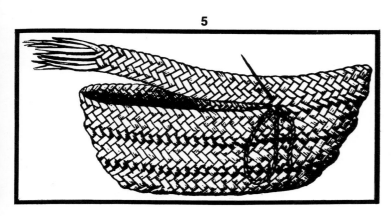

5

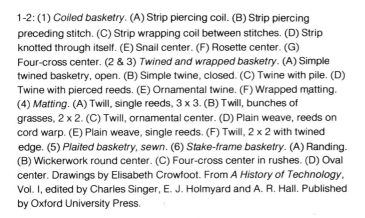

6

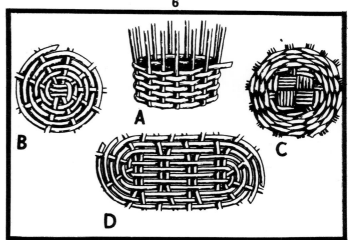

1-2: (1) *Coiled basketry*. (A) Strip piercing coil. (B) Strip piercing preceding stitch. (C) Strip wrapping coil between stitches. (D) Strip knotted through itself. (E) Snail center. (F) Rosette center. (G) Four-cross center. (2 & 3) *Twined and wrapped basketry*. (A) Simple twined basketry, open. (B) Simple twine, closed. (C) Twine with pile. (D) Twine with pierced reeds. (E) Ornamental twine. (F) Wrapped matting. (4) *Matting*. (A) Twill, single reeds, 3 x 3. (B) Twill, bunches of grasses, 2 x 2. (C) Twill, ornamental center. (D) Plain weave, reeds on cord warp. (E) Plain weave, single reeds. (F) Twill, 2 x 2 with twined edge. (5) *Plaited basketry, sewn*. (6) *Stake-frame basketry*. (A) Randing. (B) Wickerwork round center. (C) Four-cross center in rushes. (D) Oval center. Drawings by Elisabeth Crowfoot. From *A History of Technology*, Vol. I, edited by Charles Singer, E. J. Holmyard and A. R. Hall. Published by Oxford University Press.

2. *Twined* baskets are formed by intertwining two "weft" threads between parallel fibers or bunches of fibers. The technique, regarded by some as a halfway step between plaiting and weaving, was commonly used in mat making and was found in Mesolithic fish traps in Denmark.

3. In the *wrapped* technique, often classed as weaving even though it must be done with the fingers alone without the aid of a shuttle or heddles, the "weft" fibers pass over two parallel "warp" fibers, back under one, over two, and so on.

4. *Matting* is similar to weaving in the "in-and-out" technique, and some was possibly done on early looms. The threads are frequently twisted together for strength, as in two-ply yarn.

5. *Plaiting*, a technique that may have antedated basketry itself, is done in strips and then sewn together. Although subject to varying interpretations and definitions, Irene Emery in *The Primary Structures of Fabrics* characterizes plaiting as a structure made from *one set of elements* in which the elements, trending now to the right, now to the left, interlink with adjacent elements. (Others insist that only mechanical shed-making devices distinguish weaving from plaiting.) In Neolithic times plaiting was already well developed.

6. *Wickerwork*, or *stake-form*, basketry appears to be a later technique in which pliable strands are "woven" in and out of rigid stake frames.

Basketry technique, in addition to providing a vessel for carrying fish from the fish traps and roots and berries from the forests, was eventually applied to hampers, cradles, shields, quivers, and sieves. One use suggested another. Matting was used for carpets, seats, hangings, coverings, and wrappings as well as temporary shelters and houses.

The earliest evidence of basketry comes from Guitarrero Cave in Peru, c. 8600-8000 B.C., but examples almost as old have been unearthed from certain Great Basin sites, such as Danger Cave, Utah in North America. The twined basketry and bags from these early sites suggest that twining may be the oldest textile technique and the direct ancestor of weaving—though some believe that weaving was derived not from one technique but from many.

Before 6000 B.C., because of the lack of evidence of looms, distinctions between basketry and weaving are often difficult to make, but about this time it becomes clear that basketry and weaving were heading in different directions. In the early 1960s fragments of plainwoven cloth with up to 30-x-38 threads per inch—as fine as today's lightweight wools—were found at Çatal Hüyük in Anatolia and dated c. 6000 B.C. (fig. 1-3). The fiber might have been flax or possibly wool, but the threads were smooth and well prepared for weaving. The presence of a heading cord on some fragments suggests that they were woven on a warp-weighted loom (see Chapter 2), but the horizontal ground loom (see Chapter 3) was an equal possibility.

1-3: Carbonized textile fragment from Çatal Hüyük VI, c. 6000 B.C. Photograph by Arlette Mellaart.

From another layer at Guitarrero Cave a few shreds of weft-faced weaving, c. 5780 B.C., have raised the possibility that these early hunter-gatherer people possessed some kind of loom much sooner than previously imagined. Until recently it was believed that Peruvian weaving dated from the beginning of the Preceramic Period, c. 3250 B.C., when the art of basketry was already well advanced. Some claimed that even then weaving was still a minor technique used merely to repair holes in twining (fig. 1-4). The true loom with heddles probably did not appear in Peru until the Ceramic Period, c. 2200–2000 B.C., when it may have been introduced along with ceramics by new settlers from outside Peru. But more on that in Chapter 5.

In the Old World mud impressions of plain-weave basketry from Jarmo, Iraq have been dated by the carbon-14 process c. 6750 B.C. Baskets themselves have been found at the Fayum excavations in lower Egypt and at Badari on the east bank of the Nile (dating c. 5200–4600 B.C.), but by this time the art is already quite old, as the technique is highly developed.

Other examples of early weaving patterns appear on shards of pottery. For example, pot bases from Late Neolithic Hungary and the Balkans show 2-x-2-twill mat impressions. This is perhaps due to setting the pottery on woven (perhaps plaited) mats to dry, but it has also sug-

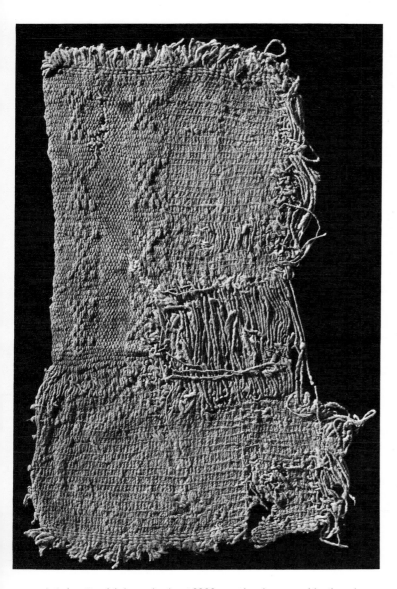

1-4: A cotton fabric made about 2300 B.C. showing a combination of weaving and twining. The warp-float figures of the woven area are repeated in the same positions on the reverse side. The wefts employed in the weave are paired for twining the balance of the warp. Huaca Prieta, Chicama Valley, Peru. Courtesy of the American Museum of Natural History.

on what a loom is—with the differences usually confined to the degree of mechanization involved. Some insist that the true loom is characterized by a mechanical shedding device that opens the entire width of the warp at once for the insertion of the shuttle. For this book, however, the term "loom" is best defined more generally as any frame or contrivance for holding warp threads parallel to permit the interlacing of the weft at right angles to form a web.

As long as the material to be woven was fairly rigid, no additional apparatus was necessary. Certain Indians of northern Canada, for example, wove blankets by laying strips of twisted rabbit skin on the ground and intertwining the weft by hand. The skins, perhaps with the aid of rocks to hold them in place, could be manipulated without tangling.

How the loom developed was to a large extent dependent on what fiber was used for the warp. The earliest looms were probably much like the Ojibway bag loom (fig. 1-6) in construction: a cord was simply stretched between two uprights (or, even simpler, the cord could have hung from a tree or have been strung between trees) from which the warp threads were freely suspended. The Ojibway warps usually consisted of silk grass, Indian hemp, the shredded bark of mulberry and cedar trees, and occasionally buffalo hair, fibers that were stiff enough to hang relatively parallel by themselves.

If, as related in the African legend above, textile development proceeded from a search for progressively finer and more pliable "bush-rope," man must have soon discovered that he needed some additional aid in holding the warp parallel. Accordingly, he probably replaced the horizontal cord with a wooden branch or beam and hung weights from the warp to keep the thread in line. This method worked well with both flax and wool. The flax fibers were long and tough—though quite short in comparison with flax—and the wool fibers had tiny barbs or scales that tended to lock the lengths together (fig. 1–7). As we shall see in subsequent chapters, a different kind of loom was invented to accommodate fine weaving with the shorter, less sticky cotton fibers and the long but delicate filaments spun by the silkworm in China.

Although the warp-weighted loom (see Chapter 2) and its linen and wool textiles are generally credited to Neolithic man in Europe, in other cultures looms may have evolved along different lines. The earliest representation of a loom, for example, dated c. 5000 B.C., illustrates a horizontal ground loom from Badari (fig. 1–8). On this loom the warp is stretched horizontally between two beams pegged a few inches above the ground. The three crossings at the left side of the loom represent picks of the weft; the three lines across the middle of the warp probably represent the sword beater, heddle rod, and shed rod. Its simplicity is such that a similar loom is still used today by Bedouin nomads; if they have to move in midweave, they simply pull up the pegs and roll up the loom—cloth, warp, heddles, batten, and all.

gested an interesting theory regarding the origins of pottery itself. Some impressions possibly were made by lining baskets with clay to make them waterproof; one of these baskets may have accidentally fallen into the fire, thus burning away the basket and hardening the pot. (In the United States netting and similar structures were used to decorate as well as to support pottery during construction, but these aids were removed before firing the clay.) Since few of the mats or baskets themselves have survived, it is to these fragments of pottery that we owe much of our knowledge of early weaves (fig. 1-5).

THE EARLIEST LOOMS

It is generally agreed that the weaving of textiles on looms began during the Neolithic Age. Experts differ, however,

1-5: Plain-weave textile impression on pottery shard from Quetta Valley, West Pakistan. Date unknown but possibly as old as 3500 B. C. Photograph by Dr. Junius B. Bird, American Museum of Natural History.

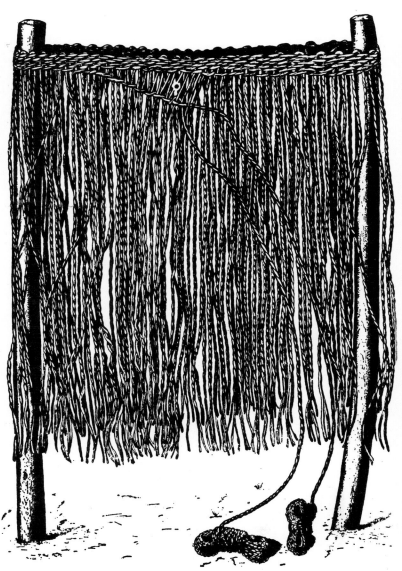

1-6: Ojibway weaving frame. Ojibway bag of twined weaving in the process of manufacture. From Clark Wissler, *The American Indian*, 1917.

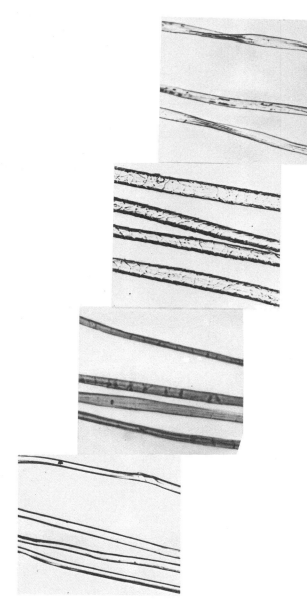

1-7: Photomicrographs of the four principal natural fibers. *From top down:* cotton, wool, linen, silk. Courtesy of E. I. duPont Co.

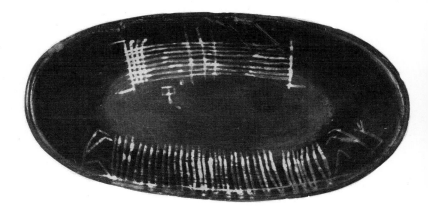

1-8: Horizontal ground loom on Badarian pottery bowl, c. 5000 B.C. It is unclear what the two figures on the opposite side of the bowl are doing; perhaps they are hanging lengths of weft over a rod in preparation for weaving. Courtesy of Petrie Museum, University College, London.

1-9: Egyptian girl in sheer linen silhouetted against a slightly enlarged photograph of equally fine ancient linen. The girl is from a wall painting in the tomb of Zeser-ka-Ra'-sonbe (c. 1420 B.C.), and the linen is from the tomb of Hat-nūfer at Thebes (c. 1500 B.C.). The Metropolitan Museum of Art, Rogers Fund, 1924.

Since the characteristics of the warp fiber influenced the development of the loom, it is useful to examine the origins of the four major yarns—linen, wool, silk, and cotton—for what they can reveal about what kinds of looms emerged where and why.

THE LOOM AND LINEN

Flax seems to have originated in the Near East and spread from there to Europe, Egypt, and other irrigated areas. But the Egyptians—perhaps because of a lack of substantial evidence elsewhere—are often referred to as the world's first skilled weavers. They certainly were cultivating flax by the fourth millenium B.C.

The flax fiber is obtained from a plant (appropriately called *Linum usitatissimum*—"most used linen") that grows from 24″ to 40″ high. Pliny, writing in the first century A.D., described a fiber-preparation system that was essentially the same as the ancient Egyptian method. It consists of tying together small bundles of ripe (yellowish) flax and hanging them to dry in the sun for several days. The stalks are then weighted in warm water (retting) until the outer coat becomes loose and they are again dried in the sun. When thoroughly dry they are pounded open on a stone (breaking). The outer skin is combed off by pulling the flax through iron spikes (hackling), and the inferior fiber nearest the skin is saved for lampwicks. The discarded skin is used for fuel. The pith, containing several grades of whiteness and softness, is then combed, separated, and spun. As thread it is soaked and repeatedly beaten before weaving and as fabric it is beaten again, for, as Pliny said, "it is always better for rough treatment."

The best record of early linen comes from the same sites as the baskets of the Fayum and Badari, dating c. 5000 B.C. These examples are all in plain weave, as was typical of all Egyptian linen until c. 2500 B.C.; the earliest piece contained 20 to 15-x-25 to 30 threads per inch. The thread itself was lightly spun and two-ply. By the First Dynasty (3250–2800 B.C.), however, the Egyptians had so mastered the art of spinning and weaving that they were producing extraordinarily fine linen, far superior to that of contemporary Europe (fig. 1–9). Textiles from the tomb of Zer at Abydos show a delicate yarn with a count of 160-x-120 threads per inch. One mummy wrapping contained an incredible 540 warp threads per inch, a feat never known to have been duplicated.

All the evidence indicates that textiles from this period were woven not on the warp-weighted loom of Neolithic Europe but on a horizontal ground loom (see Chapter 3). It is possible, though, that a warp-weighted loom was used in ancient Egypt as well. Loom weights were found at Lisht that date to the Middle Kingdom period, but the warps from which these weights were suspended may have run horizontally (see fig. 6-11), unlike their European counterpart.

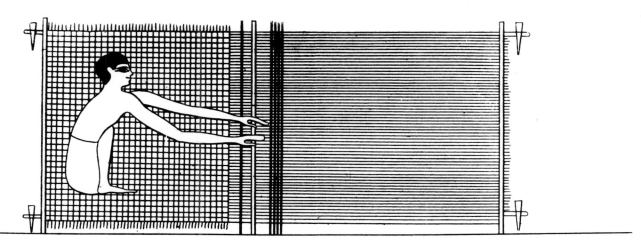

1-10: Representation of an Egyptian mat loom from the Khety tomb at Beni Hassan, c. 2000 B.C. The weaver is beating in a shot of weft. The four bars in front of his hands are probably lease sticks. After Champollion, *Monuments de l'Egypte et de la Nubie*, 1845.

On the earliest horizontal looms the weft was probably darned in by hand without the aid of shed rod or heddles, as represented in tomb drawings of the Khety mat loom (fig. 1-10). At a later stage a shed rod was probably used to open one shed, while the weft for the countershed was still darned. Heddles were finally added to form the countershed, and the "true" loom in Egypt was complete.

Since no early looms elsewhere have survived the ravages of time, moisture, insects, and fire, their existence must be inferred from the fragmentary evidence of textiles and loom weights found in archaeological ruins such as the European Lake Dwellings. The Lake Dwellers of Neolithic Switzerland and northern Italy (c. 3750 B.C.) built villages on stilts around the edges of Alpine lakes. At the earliest site at Robenhausen archaeologists have found bundles of flax fiber, fine and coarse linen thread, looped and loosely woven cloth in plain weave (fig. 1-11), and a variety of spindle whorls and loom weights of stone and pottery. These Neolithic weavers were familiar with intricate patterning that included the use of colored warp and weft threads for stripes and brocading (fig. 1-12). The discovery of loom weights and implements used in the preparation of flax fiber—such as a broken hackling board and some roughing combs—indicates that these people had been weaving for some time. (Loom weights have also been discovered in Anatolia, in Megiddo, Palestine, dated c. 3000 B.C., and in Troy, c. 2500 B.C., indicating the use of a similar loom, though perhaps for weaving wool.)

The earliest representation of the Lake Dwellers' type of loom appears on an urn from Oedenburg, Hungary from the Hallstatt period in the early Iron Age, c. 800 B.C. (fig. 1-13). The stylized picture shows two rows of warp weights, front and rear, lease sticks (or perhaps shed rods for weaving twill), a spinner with a suspended spindle, and a second weaver holding either a small frame loom or possibly an embroidery frame.

The cultivation of flax disappeared from the Lake Dwellings during the Bronze Age, probably due to climatic changes, and wool emerged as the dominant fiber for weaving.

THE LOOM AND WOOL

Not much evidence of wool fabric has survived in Neolithic Europe (wool is a highly perishable fiber), but enough sheep bones have been found to suggest that Neolithic man raised sheep. (Next to the dog, the sheep—or goat—is believed to be the earliest animal domesticated by man.) It is now well known that sheep and goats were first domesticated in Southwest Asia, with sheep in Iraq c. 8500 B.C. and goats in Iran about a thousand years later. In all probability the weaving of wool originated there as well and migrated west with the Neolithic and Bronze Age cultures to Europe.

Abraham, Jacob, and Laban were all shepherds. Abraham, originally from Mesopotamia, left the land of the two rivers in search of better grazing along the Fertile Crescent, settling eventually in Canaan. The Hebrew flocks were enormous by present standards. Moses took 675,000 sheep from the Midianites, the ancestral tribe of his wife. With the defeat of the Moabites, about 700,000 sheep were added to the Hebrew holdings.

The early Hebrew loom must have been similar to the Egyptian horizontal ground loom, though probably manipulated entirely with the fingers. It must have been on such a loom that Delilah wove Samson's hair (Judges 16:14), for she did her weaving while he slept, and he presumably slept stretched out on the ground.

17

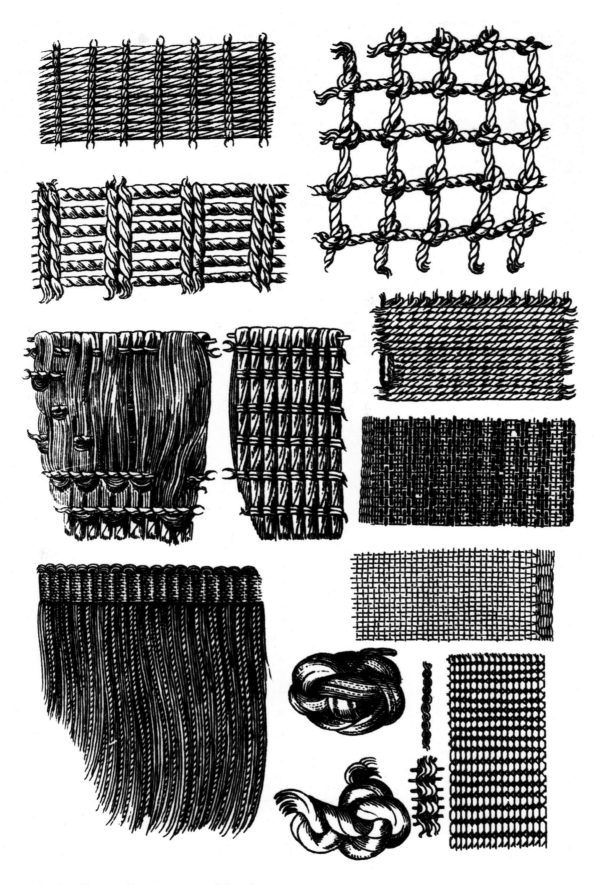

1-11: Examples of flax, bast fibers, and loosely woven cloth from the Lake Dwellings at Robenhausen and Wangen. From Ferdinand Keller, *The Lake Dwellings of Switzerland and Other Parts of Europe*, 1879.

1-12: Reconstruction of Neolithic brocaded-linen fabric from Irgenhausen. Photo: Schweiz. Landesmuseum.

1-13: Oedenburg (now Sopron, Comitate Györ-Sopron, Hungary) loom, from Hallstatt period, excavated in 1891 by Prof. Dr. Rudolf Hoernes. Museum of Natural History, Vienna.

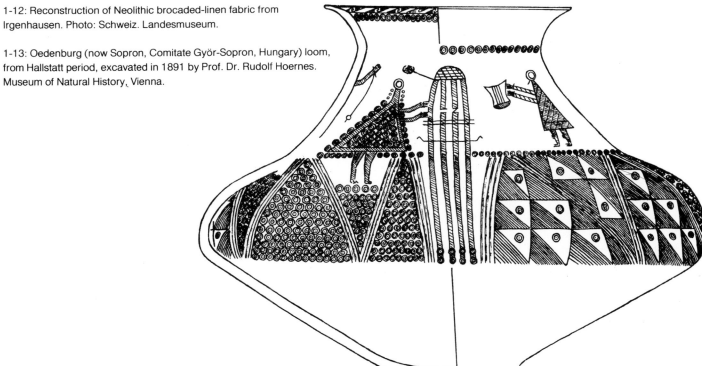

The importance of sheep in ancient times has been documented by the discovery of seals bearing rams' heads in Tepe Gawra, or Great Mound, in Mesopotamia. The seals, dating back to 4850 B.C., suggest that trade existed in wool some thirty centuries before the Fall of Troy. Further evidence of the early use of wool was found in mosaics on walls in Sumer, c. 4400 B.C., showing three different breeds of sheep. (Early sheep, unlike the thick-fleeced breeds of today, had long, coarse hair with a downy undercoat. The fleece was in fact so hairy that experts mistakenly thought for some time that the wool in certain weavings was mixed with deer hair.)

In spite of its Middle East origins the earliest extant fabric containing wool dates from the Late Neolithic Age in Europe. One scrap, in which only the weft remains, shows a mixture of sheep wool and horse, cow, and goat hairs. These early weavers probably carded their wool first with teasels and later with leather-backed cards set with thorns. Many textiles made completely from wool have come from Bronze Age graves in Scandinavia, c. 1300 B.C. Coffins of hollow oak logs found in bogs have yielded scraps of woolen cloth that seem to be an early form of tweed. Samples show a coarse, plain-weave fabric with 13-x-10 to 7-x-6 threads per inch. Some pieces had a separately woven starting border, a device used for even warp spacing and for a third selvage (see Chapter 2). The excavations indicate that these early Bronze Age weavers wove belts, tassels, caps, cloaks, and tunics, often decorated by embroidery (figs. 1-14 and 1-15). The starting borders on some pieces indicate the use of a warp-weighted loom.

THE LOOM AND SILK
The first mention of silk by a European occurs in the writings of Aristotle in the fourth century B.C., but the secret of its manufacture remained a mystery to the West until the reign of Justinian (A.D. 527–565). In A.D. 552 two Nestorian monks, former missionaries to China, fulfilled their Emperor's wishes by smuggling a few silkworm eggs and mulberry-tree seeds out of Khotan in hollow bamboo staffs. With their arrival in Constantinople, the silk industry was born in the West.

Although silk textiles were well known in the West since the trade explorations of the Han dynasty (206 B.C.–A.D. 220), before the Nestorian theft there were some curious theories regarding its manufacture. Pliny, writing in the first century A.D., believed that the Chinese combed silk from the mulberry trees. Pausanias, a traveler from Asia Minor writing in the second century A.D., came somewhat closer, believing that the fiber was produced by a strange animal that the Greeks called Ser. "Its size is twice that of the largest beetle. In other respects it resembles the spiders, which weave under the trees. It has also the same number of feet as the spider, namely eight. [The silkworm has

fourteen.] In order to breed these creatures, the Seres [the natives of Serica, the silk weavers] have houses adapted both for summer and winter. The produce of the animal is a fine thread twisted about its legs. The Seres feed it four years on 'panicum.' In the fifth year they give it green reed, of which it is so fond as to eat of it until it bursts, and after this the greatest part of the thread is found within its body."

Although the silkworm can be called gluttonous (it is said that worms from one ounce of eggs will eat in six to eight weeks a ton of mulberry leaves), Pausanias had his theory, at best, inside out. The silk filaments are not recovered from inside the silkworm but from a cocoon that the silkworm creates by means of two spinnarets. These incredible organs produce a continuous double thread 600 to 1,200 yards long, so fine that a dozen filaments must be twisted together to render them practical for weaving. The cocoons were then boiled to kill the chrysalis inside before it hatched and destroyed the threads by gnawing through them to freedom. Boiling also dissolved the ceresin that glued the filaments together, allowing the silk to be reeled and otherwise prepared for weaving (see fig. 6-13).

The delicate silk threads could not be woven on the warp-weighted loom but required the development of a more sophisticated apparatus. What the earliest Chinese looms looked like is not known, but according to one scholar, Gaines K.C. Liu, silk cloth was reportedly used during the reign of Shen Nung (traditionally dated 3218–3079 B.C.), several centuries before the discoveries of the Empress Si-Ling-Chi. In Liu's view Si-Ling-Chi perhaps represents the first governmental interest in silk weaving, but he credits the invention of the loom itself to a minister of Hoang-ti by the name of Pei Yu. As for the silk in use before this time, Liu states cryptically only that "it was woven laboriously by hand."

Fragments of twill damasks preserved on Shang period bronzes (c. 1600–1027 B.C., or about the time of early Scandinavian Bronze Age wool) suggest the use of a complicated loom, probably with many heddles. Luther Hooper asserts that silk weaving necessitated the use of treadles, and since the Chinese monopolized silk weaving until the third century A.D.—when four Chinese weavers were kidnapped and taken to Japan—he concludes that the treadle loom was probably a Chinese invention. Other authorities attribute it to India.

THE LOOM AND COTTON
The earliest cotton weaving has traditionally been associated with India, though it appears to have developed independently almost as early in Peru. From the ruins of Mohenjo-Daro in the Indus Valley of Pakistan (c. 2500 B.C.) a small piece of cloth and two lengths of yarn, one 12- and the other 24-ply, have been recovered. That they are made from cultivated, not wild, cotton suggests that weaving originated considerably earlier in this area. (The

1-14: Bronze Age man's cloak from Grave A at Borum Æshøj, Denmark. Published by permission of the Danish National Museum.

Our word "cotton" derives from the Arabic *qutun*. Like silk, the fiber was obviously a mystery to the ancient classical world, for the early Greek writers thought that it came from trees. Herodotus wrote in 445 B.C.: "The wild trees in that country [India] bear for their fruit a fleece surpassing those of sheep in beauty and quality and the natives clothe themselves in cloth made therefrom." According to Philostratus: "The material of which the Brahmans make their raiment is a wool that springs wild from the ground, white like that of the Pamphylians, though it is of a softer growth and a grease like olive oil distills from it. This is what they make their sacred vesture of, and if anyone else except these Indians tries to pluck it up, the earth refuses to surrender its wool."

The cotton fiber, having neither the scales of wool nor the length of flax (Indian cotton is 3/8″ to 1″ long), requires great skill in spinning. So much skill, in fact, that for their finest muslins the Indians would not allow the yarn to be spun by women over the age of thirty. The

1-15: Bronze Age girl's corded skirt from Egtved, Denmark. Published by permission of the Danish National Museum.

Rigveda, dated c. 1500–1000 B.C., describes Day and Night as two female weavers "intertwining the extended thread.") Other scraps have been found at the Dura-Europos site in Syria, but these are probably of Indian origin since cotton was not then grown locally. It is assumed that cotton cultivation spread west from India, reaching the Persian Gulf in ancient times. From there it may have spread to Arabia, Ethiopia, Nubia, Egypt, and, by the early Christian era, to the East as well.

suspended-spindle method would not work with cotton. Instead the spinner rested the tip of the delicate spindle in a bowl to steady it. A bit of water in the bowl gave moisture to the thread, increasing its stickiness. "The rigid, clumsy fingers of a European," commented one observer, "would scarcely be able to make a piece of canvas with the instruments which are all that an.Indian employs in making a piece of cambric."

Of one of these instruments, the early Indian cotton loom, little is known, but most authorities agree that it has probably changed little in 4,000 years. This loom is described in Chapter 6 as the pit-treadle loom—complete with heddle harnesses, reed, and treadles. It is likely, though, that a more primitive version existed, for heddle harnesses, reed, and treadles represent later developments in most weaving cultures.

Most of the evidence of early cotton in the Americas comes from Peru, which, due to an extraordinarily dry climate, offers the most complete textile history of all the ancient cultures.* Cotton was probably cultivated in the Chilca Valley of Peru c. 4750–3300 B.C., but until the Ceramic Period, beginning c. 2200–2000 B.C.—depending on the location of the site—eighty to ninety percent of fabric finds were twined, not woven. The remainder was mostly "knotless netting," or looping, with only scattered examples of plain weave.

The scarcity of plain weave suggests that the earliest Peruvian looms lacked heddles. No loom fragments survive from the Preceramic Period, but a frame loom that varied in size according to need may have been used. Some of them, judging from early fabrics, must have been as large as 4' x 5'. The fibers were manipulated by the fingers. Because of an enormous advance in culture that accompanied the Ceramic Period, some experts feel that the heddle loom did not evolve from a more primitive indigenous form but was the product of a more advanced culture that migrated into the country, perhaps from an area east of the Andes (see Chapter 5).

In North America there is little evidence of widespread weaving until much later, in part for want of a suitable fiber. Neither silk nor flax was indigenous, and sheep were not introduced until the Spanish conquests of the sixteenth century. According to Kate Peck Kent cotton—and the loom—were not introduced into what is now the United States from the south until c. 700 A.D. during the early Pueblo I period. Prior to that Indians of the United States wore the skins of the animals that they hunted. Moose hair,

*Other conditions besides aridity that have preserved textiles from ancient times include: the permafrost conditions of Norse burials in Greenland; chemical preservatives such as the tannin that preserved certain textiles in Danish Bronze Age burials; metallic salts such as those that preserved silk fibers on early Chinese bronzes; and carbonization, in which intense heat without sufficient oxygen for complete combustion rendered a textile chemically inert, as with the textile scraps found at Çatal Hüyük, c. 6000 B.C.

dog hair, and rabbit fur were sewn or twined on for decoration or warmth but generally not woven.

Three types of looms eventually made their appearance in the Southwest: the horizontal loom, pegged to the ground like the early Egyptian looms, among the Pima and Maricopa of the southern areas; the vertical, or Navajo-type, loom, favored by the Pueblo weavers; and, less popular but possibly the earliest, the backstrap, loom.

Wherever weaving began, there we must look for the origins of the loom. Along the fertile river valleys of the Hwang Ho in China, the Indus in Pakistan, the Tigris and Euphrates in Mesopotamia, the Nile in Egypt; along the river valleys of the dry coastal plain of Peru; wherever man settled down and turned his hand to agriculture and domestic chores, there we can expect to find the earliest looms. But the evidence is scanty, and much of our knowledge of early looms—when the heddle was introduced, how the reed developed, where the foot treadle originated—is merely supposition based on later developments.

The basic mechanisms of the loom evolved very early, too early to be identified by either place or time. The history of the loom is essentially a history of mechanical improvements that were directed mainly at increasing the *quantity* or *speed* of fabric production. But it would be a mistake to assume that the technological advances necessarily improved the *quality* of the woven material. In fact, a sizable number of textile historians have argued the contrary. As Hooper said, "Indeed the skill and imagination of the textile artist—as of all others—is thwarted and impaired by almost every invention which increases the speed and uniformity of production."

Each new development brought with it its own limitations. In general, the further removed the yarn became from the actual hands of the craftsman, the less flexibility he had to exercise his inventiveness and the more quality suffered. This generalization is not without exceptions, but if proof were desired, one need only look to the textiles of Peru. With the invention of the heddle, lease cord, bobbin, and batten, the Peruvian backstrap loom ceased to evolve. Yet on this primitive instrument were woven the most exquisite fabrics—using every possible weaving technique known to man—that the world has ever seen. From ancient Egypt came linen so fine that you could see the limbs through it. From early India came cotton fabric so delicate that it became invisible when laid on the dew-moistened ground, cotton spun so fine that one pound of yarn would stretch 250 miles.

The superiority of primitive weaving is not the main theme of this book. Yet in the face of man's otherwise stunning technological achievements it is worthwhile to bear in mind that it was on the handloom—often just a bundle of sticks and cords—that this heritage of magnificent textiles was woven.

2. The Warp-weighted Loom

From outside the house-gates they heard Circe, the Goddess with the comely braided hair, singing tunefully within by the great loom as she went to and fro, weaving with her shuttle such close imperishable fabric as is the wont of goddesses, some lively lustrous thing.—Homer

OPERATION

Imperishable or not, neither Circe's fabric nor any other ancient Greek weaving has yet been discovered. Our knowledge of Classical and Hellenistic textile technology derives from literary and artistic sources, as often as not from inadvertent comments such as the above. That Circe "went to and fro" as she wove may not sound like scientific evidence, but it strongly suggests the use of a warp-

weighted loom. As far as we know, the warp-weighted loom is the only known loom before which the weaver might have to stand—much less pace! In Circe's case, however, we also have evidence from painted pottery vessels indicating, however crudely, that her loom was warp-weighted. A Kabeiric scyphus from the fourth century B.C. (fig. 2-1) depicts such a loom, with Circe offering a reluctant Ulysses the potion that has transformed his companions into swine. A fifth-century B.C. Boeotian vase (fig. 2-2) shows a similar scene.

What all warp-weighted looms have in common is the technology embodied in their name—a system of holding the warp threads parallel and under tension by tying them in small bunches to weights of stone, pottery, or metal.

2-1: Circe's loom. Kabeiric scyphus with Odysseus and Circe, 4th c. B.C. Courtesy of the Ashmolean Museum, Oxford.

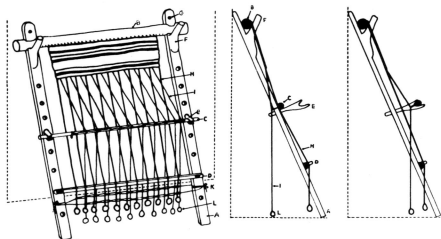

2-2: Circe's loom. Boeotian vase, c. 450–420 B.C. Reproduced by courtesy of the Trustees of the British Museum.

2-3: Simplified diagram of the warp-weighted loom. (A) Upright. (B) Beam. (C) Heddle rod. (D) Shed rod. (E) Supports for heddle rod. (F) Crotches for holding beam. (G) Hole for nailing upright to wall or beam. (H) Front warp threads. (I) Back warp threads. (K) Chained spacing cord. (L) loom weights. Norsk Folkemuseum, Bygdøy-Oslo.

2-4: Telemachus and Penelope at the loom. Representation of a painting on a Greek drinking vessel, c. 460–450 B.C., in the Chiusi Museum, Italy. From Wilhelm Kraiker, *Die Malerei der Griechen*, 1958.

How this system typically works is illustrated in the simplified diagrams (fig. 2-3). The warp is divided by taking alternate threads into the front (H) and back (I) warps, which are kept apart by a shed rod (D). Leaning the loom against a wall or rafter forms one of the two sheds needed for weaving. The second shed is formed by knitting heddles to the back warp and around a heddle rod (C), which can then be pulled forward and rested in supports (E) to open the shed. The front warp thus remains stationary, while the back warp is shifted back and forth to change sheds. Using a sword beater, the weaver beats the weft upwards against the fell of the cloth. Wool was the ideal fiber to use on the warp-weighted loom because its interlocking barbs kept the weft from slipping down as new picks were inserted.

As the weaving proceeds from the top downward, the position of the heddle rod can be lowered accordingly by moving the supports (E) down the uprights. The woven portion can also be rolled up on the top beam (B) to keep the fell of the cloth at a comfortable working height. With the innovation of a revolving top beam, it became possible to weave fabrics that were longer than the height of the loom. Both Circe's loom (figs. 2-1 and 2-2) and one representation of Penelope's loom, painted on a fifth-century B.C. scyphus from Chiusi (fig. 2-4), show a considerable portion of completed cloth rolled up on the beam, yet the weights still hang near the ground. This means that, prior to winding the cloth on the top beam, the weights must have hung from some position along the warp threads other than their ends, with the extra lengths either looped back on top of the weights or draped over an additional beam on the loom (see fig. 2-11).

The history of the warp-weighted loom is long and impressive. In Europe the loom was used during Neolithic times, and it persists to the present day in parts of Norway. Some believe that it originated even earlier in the Near East. Loom weights have been found in Çatal Hüyük, an ancient city in Anatolia that dates to the seventh millenium B.C. Other examples have been found in the first settlements of Troy (third millenium B.C.), Palestine, Crete and the Greek mainland, Neolithic Eastern Europe, the Swiss Lake Dwellings, and thereafter all over Europe, possibly even east of the Urals.

2-5: Loom weights found at Olynthus. Courtesy of The Johns Hopkins University Press.

Fig. 2-5 illustrates the varying shapes of loom weights found at Olynthus, a Grecian city that dated from Neolithic times and reached its peak with a population of over 2,000 in the fifth century B.C. Heavier weights resulted in greater warp tension and closer, firmer weaving; smoothness and evenness of the web depended on the weaver's selection of equal weights for each group of threads. The weaver could equalize the tension from unequal weights by attaching more threads to the heavier weights and fewer to the lighter ones. The weights found at Olynthus were usually made of red clay and ranged from 3/4 to 15 ounces, but by themselves loom weights reveal very little about what kind of loom supported them.

Fortunately, we have the evidence from painted pottery representations as well. The oldest of these seems to be the highly stylized drawing on the Oedenburg urn (fig. 1-13) of the Hallstatt period. Although no uprights or top beam is shown, it is clear that this loom had two rows of weights, suggesting the front-back divisions of the warp for forming sheds. The diagonal cross-hatching could be interpreted as an indication of twill weaving, which would require what might be three heddle rods transversing the warp below. But the nature of the drawing leaves much open to speculation. Almost contemporary with the Oedenburg urn is a picture on a Grecian aryballos (c. 600 B.C.) that illustrates the weaving contest between Arachne and Athena, but the detail of the looms is unclear. Equally problematic is Ovid's description of that famous contest in the *Metamorphoses,* for he portrays looms used in his own time, the Augustan Age:

No more delay: each to her corner gone,
They set their looms, and stretch the warp-threads on;
Fast to the beam the fine-spun threads are tied,
Which, parted by the reed, stand side by side;
And when the shed divides them, fingers deft
Make fly the pointed shuttle with the weft
Between the warp-threads; then they use the comb,
Deep notched with heavy teeth, to drive it home.

Even where the details are clearly painted, as on Penelope's loom (fig. 2-4), they may not be realistic. In that drawing, for example, each warp thread is individually weighted, but, as Marta Hoffmann suggests in her classic book, *The Warp-Weighted Loom,* if that were the case, the thread would unspin.

Such examples illustrate the problem of consulting art and literature for technological detail. Weaving was a commonplace activity in ancient times, and literary and artistic references did not require realistic illustration to be understood. The artistic license of the ancients has caused the modern archaeologist endless grief. The question, for example, of whether or not one of the horizontal bars crossing the midsection of the Greek loom was a heddle rod has occupied more pages in scholarly journals than the average person would care to contemplate.

THE GREEK LOOM

All in all the Greek warp-weighted loom has a distinguished literary history. It probably was this loom with which Penelope deceived her many suitors, and it may have been the loom on which Philomela wove the tragic story of her rape by Tereus, the husband of her sister Procne. Among other things the legend of Philomela tells us that the warp-weighted loom was capable of producing fine figured tapestries. This judgment is confirmed throughout ancient literature. In Euripides' *Iphigenia in Tauris* Iphigenia, banished from her homeland, laments that she will never again be seen "amid the merry whirr of looms embroidering with my shuttle a picture of Athenian Pallas and the Titans." In the *Odyssey* Homer describes a dazzling military robe draped over the hero Ulysses:

the *Odyssey*:

In the rich woof a hound, mosaic drawn,
Bore on full stretch and seized a dappled fawn;
Deep in his neck his fangs indent their hold,
They pant and struggle in the moving gold.

Aside from robes and tapestries the Greek loom was used to weave dresses, cloaks, mantles, curtains, and even rugs. Was it a "true" loom with heddles? The question has been much disputed, but the current answer appears to be yes. Before the invention of the heddle a shed rod presumably kept one shed open, while the weft for the countershed had to be darned in by hand. The heddle enabled the second shed to be formed mechanically. A minor technological advance, some would think: why all the fuss? The fuss is because the heddle was not a minor but a major technological advance that overcame the greatest problem of textile production—its tediously slow pace. Dr. Junius Bird, Curator Emeritus of the American Museum of Natural History, wrote of this invention: "In view of the importance of textiles in the lives of the majority of mankind, it is curious that the invention of the heddle is not recognized as one worthy to be ranked with, for instance, the discovery of methods for making fire. Both have played major roles in enabling man to utilize environments which otherwise would have been difficult or very discouraging. Both have an antiquity which, though by no means comparable, still remains a mystery, and both seem to have developed long after man was familiar with fire and textiles."

The evidence for the heddle in Greece comes from a combination of sources. Loom weights have been found—for example, at Troy—lying in two parallel rows with post holes at either end. Although none of the wooden

2-6: Loom weights as they have fallen from the loom. Sorte Muld II, Bornholm, Denmark, 4th–5th c. A.D. Published by permission of the Danish National Museum, Copenhagen.

supports has survived, it can be assumed that the weights fell to the ground as they had hung on the loom, probably released from their task by a fire. The two rows indicate that the loom did not stand upright, as is often believed, but slanted as in fig. 2-3. If the loom had stood vertically, the weights would have fallen in one row, even if separated by a shed rod. Fig. 2-6 shows a similar discovery in Denmark from the fourth to fifth century A.D. A slanting loom suggests the use of a heddle rod. Without it the weft would have to be darned in one direction, a task immeasurably complicated by separating the warp threads into two planes.

Most of the representations on pottery support this interpretation by showing the loom weights divided into two rows (e.g., fig. 2-1), with the back weights on top of the front ones. A notable exception occurs on a lecythus dated c. 560 B.C. (fig. 2-7), which shows a loom with only one row of weights. Perhaps this loom stood perpendicular to the floor, possibly constructed for tapestry work where a shedding device was a marginal advantage. It is also possible, as Hoffmann suggests, that this painting merely represents an earlier, less developed loom. The evidence is inconclusive.

Even more controversial is the argument for heddles based on illustrations of the supposed heddles themselves. The usual interpretation states that the thicker of the two crossing rods is a shed rod, and the thinner the heddle. (Some think that the two are laze rods.) Oversimplification by the artists has caused considerable disagreement. The lower bar of the above loom, for example, has been identified as a shed rod by some, but others believe that the x's

indicate that it was a warp spacer, a device like the chained spacing cord (K) in fig. 2-3, used to keep the warp threads in order. The thin rods just above it have been called variously shed rods, a single rod with a gap in it where the painter's brush ran dry, and heddle rods. If that were not confusion enough, the fact remains that if the loom were a tapestry loom, it might not have needed heddles at all. Luther Hooper, in an article in *Burlington Magazine* on "The Technique of Greek and Roman Weaving," in fact had used the freedom of Greek tapestry design as the basis for his argument against the Greek use of the heddle: if the magnificent sixteenth-century tapestries were produced without heddles, why suppose that the Greeks had them? The most recent interpretations, however, contradict Hooper and others and assert that the heddle was well known in ancient Greece. Hoffmann goes even further, believing that the European warp-weighted loom *always* had heddles: "No trace of any warp-weighted loom without heddles and heddle rods has ever been found in Western or Northern Europe. The preserved fragments of cloth provide indisputable proof that the loom, from its very first appearance, was a true loom, with heddles." (Some authorities doubt that one could discern the use of heddles from the finished product.)

Weaving on the warp-weighted loom began to decline in Greece during the first century A.D., but the loom continued in use for certain ceremonial garments until as late as the seventh century. It was replaced by either the horizontal ground loom (see Chapter 3) or the horizontal treadle loom (see Chapter 6). In northern Europe, however, the warp-weighted loom lasted considerably longer.

2-7: Warp-weighted loom on a black-figured Greek lecythus, c. 560 B.C. This detail from the vase (height 6¾″) shows women working at the loom. Note the pyramidal loom weights with rings (see fig. 2-8). The Metropolitan Museum of Art, Fletcher Fund, 1931.

2-8: Loom weight with bronze ring, as on the Greek lecythus in fig. 2-7. The figure stamped into the red-clay weight in low relief is possibly a spinner. Possibly of Italian origin. Reproduced by courtesy of the Trustees of the British Museum.

THE SCANDINAVIAN LOOM

Literary references to the warp-weighted loom are by no means confined to the classical world. In *The Story of Burnt Njal*, an Icelandic saga of the eleventh century, some women are preparing a grisly fabric of war with a weft of human entrails woven into a warp weighted with the heads of slain warriors. Fortunately, our knowledge of the Scandinavian loom doesn't depend solely on such literary clues. The warp-weighted loom was commonly used in Iceland until the late eighteenth century. It lasted somewhat longer in the Faerøe Islands and can still be found today in parts of western Norway, where the sword beaters of whalebone are similar to those used in Viking times. The evidence from the North is thus fairly recent and includes not just representations of looms but parts of looms or the looms themselves.

While there is no single Scandinavian loom, the warp-weighted looms of the Faerøes, Iceland, and Lapland are similar enough to permit certain generalizations. If we make allowances for regional variations, we can say that the ''Scandinavian loom'' resembles the Greek loom in principle. In fig. 2-9, probably a typical example from

Iceland, the uprights lean to the rear; the warp is divided by a shed rod into front and back warps; heddles are attached to the heddle rod and back warp (here pulled forward); and the top beam revolves to take up the finished cloth. The spokes at the end of the beam lodge under a crossbeam to keep the cloth from unrolling, and a chained spacing cord, transversing the warp above the weights, helps keep the threads in order Note also the pin beater inserted in the warp above the heddle rod. The loom of the Faerøes (fig. 2-10) is assumed to have been similarly constructed because the colonists from these two lands shared a common origin in western Norway. Lapland offers its own version, which again is essentially similar to that described above. It is thought that this is the same loom that the Lake Dwellers used during Neolithic times.

The earliest evidence of Scandinavian weaving comes from the Bronze Age bog finds of woolen garments in hollowed-out log coffins. The Early Bronze Age remains are usually coarse plain weaves of wool from long-haired, primitive sheep; the number of warp and weft threads per inch varies from 13 x 10 to 7 x 6. By the Late Bronze Age twills appear with some regularity; the earliest is judged to

2-9: Icelandic warp-weighted loom, reconstructed and set up before 1914. National Museum of Iceland. Photo: Gísli Gestsson.

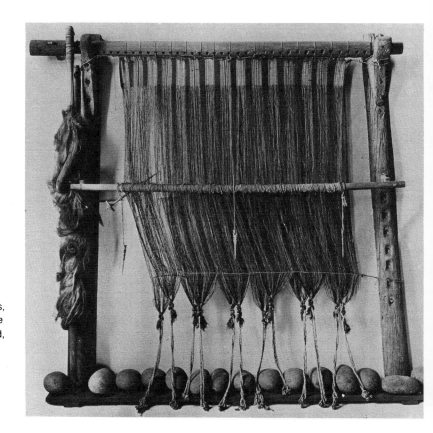

2-10: Warp-weighted loom from the Faerøe Islands, Denmark, perhaps the oldest preserved loom in the North. The warps are spaced along a heading cord, which is sewn to the top beam. Photograph courtesy of the Danish National Museum, Copenhagen.

2-11: Icelandic warp-weighted loom. Drawing by Soemundur
Magnusson Holm, 1778. (1) Uprights. (2) Crotches. (3) *Backtree*. (4)
Beam. (5) Spokes. (6) Pin beater. (7) Shed rod. (8) Heddle rods. (9)
Loku*þ*ollar. (10) Broad fixed-shed rod. (11) Temple. (12) Not shown.
(13) Loom weights. (14) Group of ends. (15) Chained spacing cord.
(16) Warp. (17) Butterfly. (18) Warp. (19) Weft. (20) Woven cloth
(*wadmal*). (21) *Meimernar*. (22) Sword beater. Det Kongelige Bibliotek,
Copenhagen.

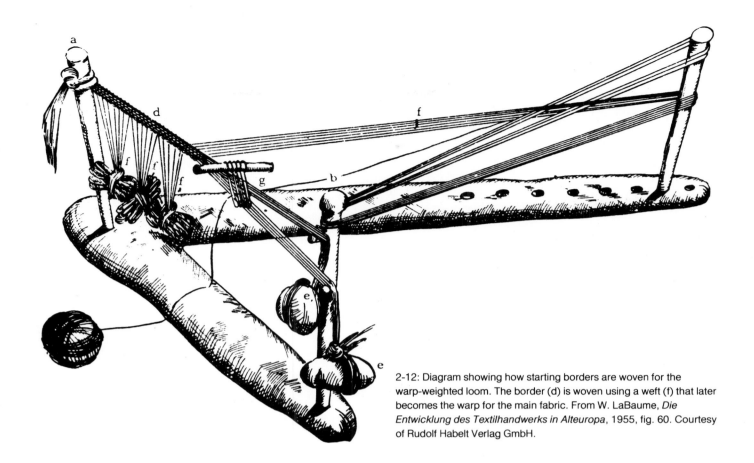

2-12: Diagram showing how starting borders are woven for the warp-weighted loom. The border (d) is woven using a weft (f) that later becomes the warp for the main fabric. From W. LaBaume, *Die Entwicklung des Textilhandwerks in Alteuropa*, 1955, fig. 60. Courtesy of Rudolf Habelt Verlag GmbH.

be a cloak of some 100″ in length from Gerumsberg, Sweden, dated somewhat earlier than the Oedenberg urn.

Although it is possible to weave a twill by "darning" the weft across the warp, the more likely method, even for these early twills, is to employ multiple heddles. At least three plus a shed rod are needed for a 2/2 twill. (Margrethe Hald, an authority on ancient Scandinavian textiles, states that "it was common, but not easy" to weave with three heddle rods on the warp-weighted loom.) This multiheddle arrangement can be seen in what is currently regarded as the oldest existing illustration of the Icelandic loom (fig. 2-11), drawn by S.M. Holm in 1778.

That the loom is represented as vertical is probably a simplification by the artist, for the Scandinavian loom is usually slanted to the rear. (Hoffmann calls Holm's drawing somewhat mischievous and misleading.) A *backtree* (3) has been added to hold the extra warp lengths. The three heddle rods (8) are not supported by brackets fastened to the uprights but seem to lean against the *meimernar*, slanting rods that appear to rest against the shed rod (10).

What neither Holm's drawing nor fig. 2-10 shows is the unusual method of warping so closely associated with the warp-weighted loom—the starting border. This border, the top edge of the fabric, was woven separately, as illustrated in fig. 2-12. Upon completion the right selvage of the

border (d) is sewn to holes along the top beam or to a heading cord; the long weft loops (f) of the border accordingly become the warp of the fabric. This may seem like an unnecessarily elaborate way to do it, but the starting border performs two useful functions: it spaces the warp evenly and provides a strong third selvage for the fabric.

There are various ways in which the starting border can be woven. Fig. 2-13 shows a Lapp warping device with pegs arranged to obtain what will become equal warp loops of a specified length. The shedding is accomplished here by means of a rigid heddle, but an older method, characteristic of the Bronze Age, was to use cards or tablets (fig. 2-14). The origins of tablet weaving are unknown, though it has been found almost everywhere in the Old World from Japan to North Africa to Iceland. Tablet weaving was by no means confined to starting borders on longer garments: it was also favored for headbands, girdles, and other narrow bands of fabric. While authorities disagree on the origins of tablet weaving, one scholar places it in Egypt before 4000 B.C. The earliest extant evidence consists of three woven bands, possibly tablet-woven, that date to the Twenty-second Dynasty (945–745 B.C.). In Scandinavia no evidence has been found that predates the Bronze Age, c. 1300 B.C.

The tablets, usually square and, in the North, made of

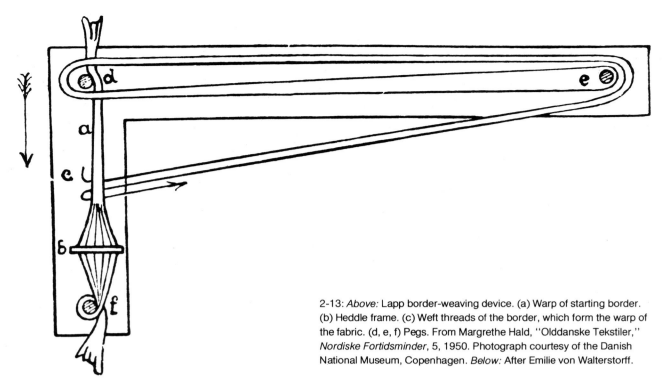

2-13: *Above:* Lapp border-weaving device. (a) Warp of starting border. (b) Heddle frame. (c) Weft threads of the border, which form the warp of the fabric. (d, e, f) Pegs. From Margrethe Hald, ''Olddanske Tekstiler,'' *Nordiske Fortidsminder*, 5, 1950. Photograph courtesy of the Danish National Museum, Copenhagen. *Below:* After Emilie von Walterstorff.

2-14: Weaving implement found in the tomb of Queen Asa, consisting of 52 wooden tablets. This set belongs to the Oseberg finds (Oslo fjord) dating from the 9th c. A.D. Oslo University, Collection of Antiquities.

bone, wood, or horn, were perforated and threaded through the corners. The number of tablets (and the weight of the fiber) determined the width of the border; the number of holes conditioned the variety of colors or patterns that could be woven. A natural shed was formed by the distance between the holes (fig. 2-15). When the tablets were given a quarter turn, a different shed was created. If the tablets were then turned back to the original position, plain weave resulted; but if they were continually turned in the same direction, the result was a corded weave.

Some authorities believe that many of the early starting borders were woven as shown in fig. 2-16. Instead of pulling the entire ball of yarn back and forth through the shed, the ball was kept to one side, and loops, or double threads, were pulled through. Thus, before beginning a plain weave the double threads would have to be crossed (fig. 2-17) to organize them alternately in the front and rear.

The loops at the opposite end of the warp must have presented certain problems. If the loom that accommodated them resembled the example in fig. 2-11, the loops must have been cut open to avoid shedding difficulties. Yet fabrics have been discovered with the looped ends intact, which suggests two other loom possibilities. In one, which Margrethe Hald describes as a

tubular loom (though it is the fabric and not the loom that is tubular—see fig. 3-16), the warp is wound continuously over two beams and a cord that is stretched between the uprights. The cord is the seam (see fig. 3-17) that joins the warp ''ends''; it can be removed, separating the ''ends, ''if a flat piece of cloth is desired. The second possibility is that the fabrics with uncut loops were woven on a simpler loom with no shedding device, such as that illustrated on the Greek lecythus in fig. 2-7 or that used by the Chikat Indians in North America (see the next section).

The width of some of the Scandinavian fabrics (9′ to 10′) suggests that two people wove on the loom at once. The procedures for weaving varied, depending on the place and on whether or not a starting border was used, but in general they were as follows. After weaving the starting border or winding a warp on pegs knocked into a door frame or set into the loom uprights, the warp was placed on the loom. It could have been either sewn directly to the top beam or draped over a heading cord that was then fastened to the beam. The warps were next separated into front and back and weighted on either side of the shed bar. The heddles were knitted on, and cords were chained across the front and back warps to keep the spacing even.

The weft was wound into butterflies for passage through

33

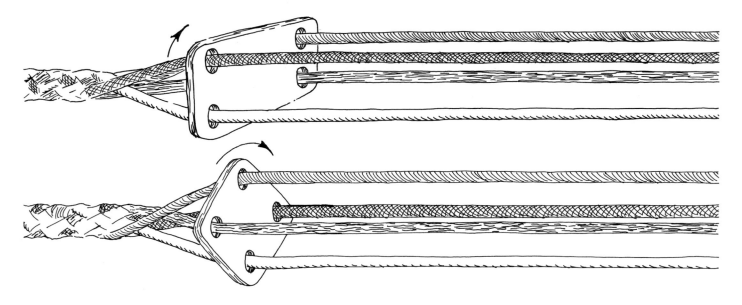

2-15: Shed formed by twisting tablets on a tablet loom.

2-16: Lapp woman weaving a starting border. The loop from the weft ball on the floor has been passed through the shed and is going to be taken around the pegs. Norsk Folkemuseum, Bygdøy-Oslo.

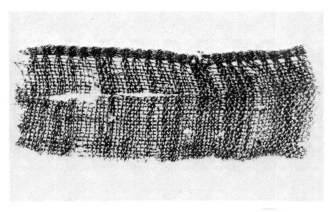

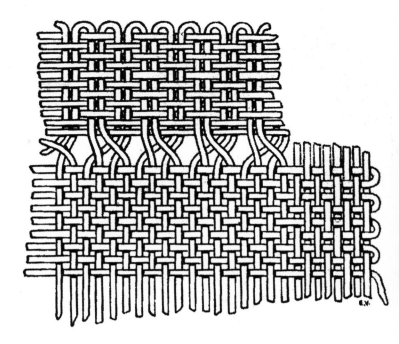

2-17: Photograph and diagram of a woven starting border from Robenhausen. Swiss National Museum, Zurich.

the warp. (The bobbin, or needle shuttle, seems to have been a late development in Scandinavia, perhaps borrowed from the horizontal loom.) In weaving the weft was thrown through several sheds before it was beaten up. The sword beaters were usually shorter than the width of the fabric and had to be inserted in several places along the fell of the cloth to compress the entire weft. These details varied widely. In Lapland, for example, no sword beater appears to have been used at all; the weft was pushed up with the edge of the weaver's hand.

As cumbersome as this sounds, the warp-weighted loom offered one distinct advantage over later looms in which the warp was stretched between two beams: when the heddle rod was pulled to change sheds, the weights rose but the tension remained unchanged. That advantage was not enough, however, to ensure the loom's survival, for its disadvantages were also distinct: the weaver had to stand at his loom; it was hard work to beat the weft upward; gravity itself tended to loosen the web (unlike the conventional tapestry loom in which the weft is beaten *down*). Nevertheless, closely woven, fine cloth was woven on the warp-weighted loom, and it survived in Europe until the horizontal treadle loom, introduced during the early Middle Ages, began to replace it. If it continued in use after that, it probably served only as a warping board for the treadle loom. In Scandinavia, however, it persisted, resisting change, into the mid- to late eighteenth century.

THE CHILKAT LOOM

Strictly speaking, the Chilkat loom (fig. 2-18) is not a warp-weighted loom but a free-warp loom: it lacks proper loom weights. Nor does it need them, because the warp is rigid enough to stay in place by itself. This stiffness is achieved by wrapping a twisted core of red- or yellow-cedar bark in mountain goat's wool. What appear to be loom weights are merely the remainder of the warp ends tied up in sacks made of goat intestines—a simple device to keep the ends clean and untangled (fig. 2-19). In spite of the lack of weights the Chilkat loom has been included in this chapter because it shares other significant attributes with the warp-weighted loom.

The Chilkat Indians of the Northwest Coast of North America were not the only New World weavers to use the warp-weighted type of loom. Similar looms were used by some of their neighbors bordering them to the south: the Haida, Tsimshian, Kwakiutl, and probably the Nootkan and Bella Coola tribes. The Chilkat examples, however, remain the best known.

The loom itself, as George T. Emmons described it in his authoritative study, "The Chilkat Blanket," "consists of two uprights resting in heavy shoes, one broad crosspiece on which the blanket is hung, and two narrow slats to keep the uprights from spreading, to hold the cover down, and to hang the extra woof on."

2-18: Chilkat blanket loom, Alaska. Traditional designs are drawn on wooden pattern boards (to the right of the loom). Courtesy of the American Museum of Natural History.

2-19: Chilkat loom with partly woven legging attached. 18″ wide at top. The Museum of the American Indian, Heye Foundation.

To prepare the loom, the warp is cut into strips and hung over a heading cord of hide (fig. 2-20), which is then laced to the top beam. A measuring stick with graduated notches (fig. 2-21) aids the weaver in cutting the warp lengths to form the characteristic V-shaped bottom of his blankets. No shedding devices are used, as the fabrics are not woven but *twined*. A twilled twining technique, called twining on alternate pairs, is used (see fig. 2-20) in which the wefts enclose two warps at a time, with each row splitting the pairs of the row above.

Unlike the European warp-weighted-loom weaver, the Chilkat sits at her loom, her knees tucked to her chin. Using only her fingers for tools, she twines in her weft of mountain goat's wool, dyed black, yellow, or bluish green or left white. Instead of working from selvage to selvage, constantly shifting her position across the loom, she sits still and weaves in vertical strips. The strips are sewn together as the work advances with a cord of wool-and-bark or sinew. As the cloth accumulates, it is rolled up onto the top beam. There the goat performs a final service by providing an intestinal cover for the cloth. Weaving a blanket may

take over a year, and the weaver may dismantle her loom and move many times during that period. The cover protects the cloth until the whole is completed.

Besides blankets the products of the Chilkat loom include aprons, sleeveless shirts, and leggings, all woven in a similar manner, following detailed patterns drawn on wooden pattern boards (fig. 2-18). The origin of the Chilkat loom is unknown. According to one Chilkat tradition, the technique was borrowed from the Tsimshian, but where the Tsimshian learned it poses the same problem. It might have derived from suspended-warp basket weaving, but no one can say with any certainty. Twining, however, is one of the most common Preconquest textile techniques in the New World and one of the easiest ways to do it is on a warp suspended from a cord.

Until recently the Chilkats continued to weave in the traditional manner, their only concession to the twentieth century being the use of commercial dyes. Their weaving, however, is no longer widespread, and the Northwest Coast culture in general has been engulfed by that of the United States and Canada.

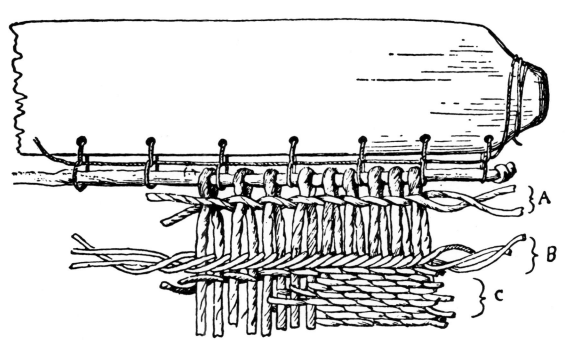

2-20: Batten and weave forming upper border of Chilkat blanket. (A) Heading cord twined around each double warp. (B) Twined three-stranded border around entire woven design. (C) Twilled twining of warp pairs for design. Courtesy of the American Museum of Natural History.

2-21: Chilkat measuring staff, length 61″. Courtesy of the American Museum of Natural History.

3. The Two-bar Loom

Great genius of the ancient times!
A loom like thine was well worth leaving;
To thee, what are our feeble rhymes?
First master of the art of weaving!

. . .

Thou breath'd the freest air of heaven,
The sun, unclouded, gave thee light;
No lamp, nor gas to thee was given;
Through day thou worked and slept at night!
—Brien Dhu O'Farrell

The indoor weaver is in a worse position than
any woman. His knees are drawn up to his heart.
He never tastes fresh air. If he does not
produce enough as the result of a day's work,
he is beaten like the lotus in the pond. He
gives bread to the door-keepers that he may
see the light of day.—Papyrus Anastasi

THE HORIZONTAL LOOM

The idea of stretching the warp between two bars is so fundamental to the weaving process that it occurs with various modifications in virtually all cultures that weave cloth. The two-bar system is found on treadle looms, tapestry looms, and backstrap looms alike, and in one or more of its various incarnations it has become the best way to hold warp threads parallel. (Lest it be thought that parallel bars are the only way to stretch a warp on a horizontal loom, see the illustration of the Syrian pit loom in fig. 6-12.) Most of what comprises a loom is in fact some kind of apparatus to support and separate the warp and cloth beams. Only gradually, as innovations occurred, did the loom frame take on the additional function of supporting other labor-saving devices, such as heddle harnesses and a reed beater.

It has often been said that the Egyptians were the first masters of the art of weaving, but the earliest evidence of woven cloth—with smooth fibers, as fine as today's lightweight wools—comes not from Egypt but from Çatal Hüyük in Anatolia, c. 6000 B.C. It is possible that these fragments were woven on a warp-weighted loom, but the evidence equally favors a two-bar loom known as a horizontal ground loom.

It is this loom that is depicted on a Badarian pottery dish (fig. 1-8), c. 5000 B.C., and that is represented on the earliest tomb paintings at Beni Hasan (fig. 3-1), c. 2000–1785 B.C., from the Middle Kingdom. A similar loom is also in use today among the Bedouin nomads of the Libyan desert. That the horizontal ground loom has persisted relatively unchanged for some eight thousand years, during which time the Sahara itself has been transformed from a lush prairie into a vast wasteland, speaks well for its serviceability to its task.

The earliest two-bar looms probably had no heddle arrangement at all, leaving the weft to be darned across, as with the Khety mat loom. Champollion's drawing (fig. 1-10) shows the weaver sitting on the woven portion as he darns in the weft strip by strip. The four lines crossing the warp in front of the weaver could represent lease sticks, but, since the drawing is purely schematic, one can't be sure. The leap from this mat loom with its reed or papyrus weft to the horizontal ground loom with linen woven "so fine in texture that a whole length could be drawn through a finger ring" was so great that it must have included several intermediate stages of development. However, when pictorial evidence of the ground loom first appeared in Badari, both the loom and the textile remains were in advanced stages of evolution.

THE EGYPTIAN LOOM

No ancient Egyptian looms have survived, but tomb paintings and one rather remarkable model, also preserved in a tomb, have given us a substantial idea of how they worked. The tomb of Chnem-hotep at Beni Hasan has provided what is probably the clearest illustration (fig. 3-1). It is also the most frequently reproduced drawing, though the details vary slightly in many of the early hand-drawn reproductions. The loom appears vertical, but in fact the warp is stretched horizontally between two beams held in place by pegs pounded into the ground. At the far end a cord has been chained across the warp to keep the threads in order. Some cloth appears to be wound around the cloth beam,

3-1: Tempera copy of a wall painting of women weaving and spinning from the Tomb of Chnem-hotep, Dynasty XII, c. 2000–1785 B.C. The Metropolitan Museum of Art.

though it is not clear how it is kept from unrolling. The loops on the left side of the cloth are a weft fringe, indicating that the fabric had only one selvage. (Louisa Bellinger, an expert on early Egyptian and Near Eastern textiles, noted that the fringe may have resulted from inserting extra weft threads at the edge to equalize unequal beating in. Since two women handled the beater and since the one who pulled could exert greater force than the one who pushed, the weft tended to compress more along the left selvage. The fringe of extra weft threads was thus added to compensate.) The weaver on the right holds the sword beater and appears to be beating in the last shot of weft. The weaver on the left, who seems to have two left hands, is holding what is probably the heddle rod with one of them and is resting the other on what might be a heddle jack, a support for the heddle rod. The rod beneath the heddle rod could be a shed rod, but if so it would more likely be placed behind the heddle rod.

The woman standing behind the weaver is generally taken to be the taskmistress, while the two women to the right are preparing the flax and spinning it. The Egyptians were superb spinners and could, as the woman here demonstrates, manage two spindles at once. (She is holding the second one behind her back.)* One commentator has stated that "they are obliged to balance themselves on a stool, and even take off most of their clothes for fear that the threads should get entangled." By standing on a stool (fig. 3-2) the spinner could spin a longer length of yarn before winding it on the spindle. (The longer the distance over which the twist is distributed, the smoother the resulting yarn.)

During the affluent Middle Kingdon (c. 2134–1786 B.C.) a middle class of shopkeepers and artisans emerged in Egypt, some of whom grew wealthy enough to erect tombs of their own. A weaving shop might have resembled the funerary model of a shop (fig. 3-3) from the estate of

*The Egyptians were not alone in their spinning skills. Herodotus (V, 12) tells the story of King Darius of Persia, who saw a Phoenician woman spinning while leading a horse to a well and carrying a pitcher on her head. He sent spies to follow her, and they reported that she filled the pitcher, watered the horse, and returned, dragging the horse on her arm, without ceasing to spin.

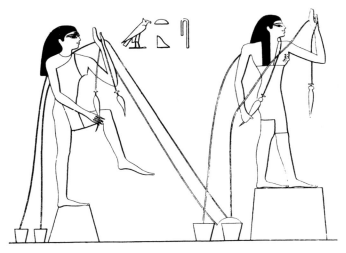

3-2: Egyptian spinners spinning two spindles simultaneously. They stand on blocks to allow for a longer and more evenly spun thread. From the tomb of Rotei, Beni Hasan, El Qadim, c. 2000 B.C. After Champollion, *Monuments de l'Egypte et de la Nubie, 1845.*

3-3: Model of a weaving shop, Tomb of Meket-Rē, Egypt, Dynasty XI, c. 2000 B.C. Photograph by Egyptian Expedition, The Metropolitan Museum of Art.

Meket-Rē, a Theban nobleman who died c. 2000 B.C. At this time and until the New Kingdom a few centuries later weaving was women's work.

At the wall to the left the three seated women are preparing the roving for spinning. The roves were rolled out on a bared knee, wound into loose balls, and placed in the ceramic pots next to the spinners. They in turn wound the roves around spindles held in the left hand (the distaff didn't appear in Egypt until Roman times) and drew them from there to the spindle in the right hand. This latter spindle would be spun on a raised knee and dropped to draw out and twist the yarn. To the right two other women are warping the spun flax by winding it in a bent figure eight around three pegs in the wall. As seen in the painting from Chnem-hotep's tomb (fig. 3-1), the looms require two weavers at the cloth-beam end, but here the two sets of weavers share an assistant who tends the warp beams of both, letting out extra warp as needed. Both beams on this loom could be rotated.

Fig. 3-4 illustrates how the shed was changed. The heddle jacks stood about a foot high and were held tightly in place by the tension of the warp—so tightly that the assistant weaver used a stone to knock them over. They had to be reset after every second shot of weft, but the skilled Egyptians, having mastered the technique, could probably do it easily.

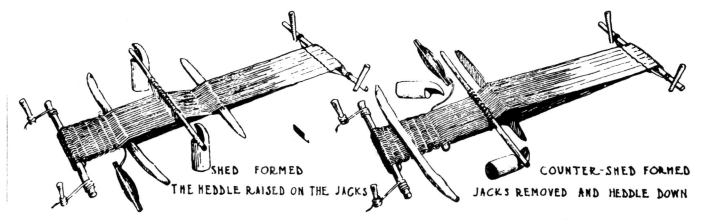

SHED FORMED
THE HEDDLE RAISED ON THE JACKS

COUNTER-SHED FORMED
JACKS REMOVED AND HEDDLE DOWN

3-4: Working model of Egyptian horizontal loom showing use of heddle jacks. Petrie Museum, University College, London.

Whether it was used for mummy wrappings or tunics, the cloth from Middle Kingdom, Old Kingdom, and Predynastic looms was tabby-woven white linen. Since the sleekness of linen yarn allowed the threads to sit closely together, most of the woven fabric was warp-rep—that is, more warp threads to the inch than weft. The use of color had to await the New Kingdom (c. 1570–1085 B.C.) and the influence of the Middle Eastern wool weavers.

THE MODERN GROUND LOOM
The modern Middle Eastern ground loom (fig. 3-5) appears to be a direct descendant of the Middle Kingdom Egyptian prototype. Grace M. Crowfoot, a textile historian well known for her work in the Near East, gave the following description (paraphrased here) of a weaver setting up a ground loom in 1921 in the Sudan. With a *Bismillah,* or word or two of charm or prayer, the weaver begins by finding a shady spot in which to locate her loom. She hammers two pegs in the ground and behind the pegs places a long stick, which forms the warp beam. Measuring the desired length of the piece to be woven with her hand, she pounds two more pegs in the ground and lays the

3-5: Bedouin ground loom in Samu'a, southern Judean Hills, used for rug weaving. Photograph by Shelagh Weir. Reproduced by courtesy of the Trustees of the British Museum.

breast beam, another long stick, behind them. With another *Bismillah* she ties strong two-ply wool to the warp beam and passes it under the breast beam and back over the warp beam, forming a continuous warp. If the warp is long, two workers lay it in by tossing the ball of warp yarn back and forth. A shed stick is inserted between the upper and lower threads close to the warp beam, where it remains until the piece is woven. Another word of prayer and the weaver sets a rod on stones on either side of the warp and laces it continuously to the lower warp threads to form the heddles.

To weave, the worker lifts the heddle with her right hand, raising the lower threads, beats them with her left hand on the warp-beam side of the heddle until the upper threads separate and fall below them, and pulls the upper ones down on the breast-beam side of the heddle rod. She then passes the weft through the space formed when the upper and lower warp threads change position. To form the countershed, she lifts the shed stick to raise the depressed threads. At the same time she pulls the lower warp threads back down with her hand, first at the back of the heddle rod and then in front. No shuttle is used; the thread is wound in a ball, sometimes with a stick in the middle, and inserted in the shed with the fingers. This arduous process is considerably more tedious with wool than with linen because of the scales on the wool fibers, which tend to lock them together.

A close relative of the Sudanese loom is the fixed-heddle loom of the Palestinians or North African Bedouins. It resembles the Sudanese loom in all respects except that, instead of lifting the heddle rod to make the shed, the rod is fixed and the warp itself is moved up or down (fig. 3-6). Unlike the Egyptian, who sat alongside the selvage, the Bedouin typically sits on top of the woven portion, pushing the heddle along in front of her as she approaches the warp beam. The beams do not revolve, and the loom is just as long as the desired length of cloth and as wide as the desired width.

Depending on the width of the loom, up to three women will squat side by side working in unison. The Bedouins use wool for their fabrics, and the sheds are changed, as on the Sudanese loom, only with great effort. The yarn is warped directly onto the beams in one long, continuous spiral, and the threads are aligned not by a figure eight in the warping but by inserting a shed stick through alternate warps as the yarn is wound. Some typical implements used in Bedouin weaving are shown in fig. 3-7.

In northern Cameroon the Doayos people weave on a loom that is very possibly an adaptation of the Bedouin loom to the north (fig. 3-8). The loom has been called an "aberrant type" of the area because it does not conform to the traditional West African models (see Chapter 6). The weaver has raised the warp several inches off the ground and, unlike the Bedouin, does not sit on the web but alongside it. The heddle rod is supported by a pair of tripods fashioned from the natural growth of tree branches (fig. 3-9) and is shifted along the warp as weaving progresses. Too thick for clothes, the strips are woven for a dowry and, later, to wrap corpses for burial—the richer the man, the more the wraps.

THE AMERICAN SOUTHWEST LOOM
The greatest virtue of the horizontal ground loom was its simplicity. It could be set up anywhere with just a bundle of sticks and pegs and perhaps a couple of stones to support the heddle rod. So primitive was this apparatus that the

3-6: Opening the shed on a ground loom with a fixed heddle rod, showing the position of warp threads on the ground loom when (a) the shed is formed and (b) the countershed is formed. Note the position of the sword beater in (b). The weaver formed the countershed (b) by reaching over the heddle rod and pulling up handfuls of the warp layer that passed over the shed rod as she pushed down on the warp layer that went through the heddles. Reproduced by courtesy of the Trustees of the British Museum.

3-7: (a) Iron hook for beating in on the Bedouin ground loom. Length 6¼″. (b) Sword beater (length 50″) and stick spool (length 42″) used on the Bedouin ground loom. Photographs by Shelagh Weir. Reproduced by courtesy of the Trustees of the British Museum.

3-8: Horizontal ground loom used by the Doayos tribe in northern Cameroon. Photograph by René Gardi.

3-9: Close up of the fixed-heddle arrangement on a Doayos loom. Photograph by René Gardi.

casual observer might easily mistake it for something else—as did Major Emory of the U.S. Army when he encountered Indian weavers in New Mexico in 1846: "A woman was seated on the ground under the shade of a cottonwood. Her left leg was tucked under her and her foot turned sole upward; between her big toe and the next was a spindle about eighteen inches long, with a single fly [whorl] of four or six inches. Ever and anon she gave it a twist in a dextrous manner, and at its end was drawn a coarse cotton thread. This was their spinning jenny. Led on by this primitive display, I asked for their loom by pointing to the thread and then to the blanket girded about the woman's loins. A fellow stretched in the dust, sunning himself, rose leisurely and untied a bundle which I had supposed to be a bow and arrow. This little package, with four stakes in the ground, was the loom. He stretched his cloth and commenced the process of weaving."

Weaving in the Southwest is usually associated with the Navajo or Pueblo vertical loom in northern New Mexico and Arizona (see Chapter 4), but the horizontal ground loom, though less celebrated, was not uncommon in the southern parts of those states. The Indians in Emory's description, Pimas, were only one of several tribes in the American Southwest that used the horizontal ground loom. Beginning about A.D. 700, and perhaps even earlier, with the introduction of cotton into the Southwest and continuing through the nineteenth century, this type of loom was used variously by the Maricopa, Papago, Opata, and Pima tribes. It is thought that the horizontal loom spread northward from Mexico into this region long before the time of the Aztecs.

Though the ground looms of these southern tribes varied in detail, they were essentially the same. An area the exact size of the cloth to be woven was first staked out. The ground was then prepared to protect the underside of the fabric, either by spreading a layer of clean sand over it or by sprinkling it with water, pounding it down, and sweeping it clean. The end beams, made of the tough inner wood of the giant saguaro cactus, were lashed to the stakes about six inches above the ground. The warp was wound over the beams in a figure eight so that the threads were kept in order, and sticks were inserted in the loops to hold the cross. Heading cords were twined about the warp threads at both ends of the loom, locking the threads in place. Each end beam was then removed and placed outside the heading cord, which was lashed firmly to it. The beams were refastened to the stakes and made taut for weaving. The heddles were prepared from one continuous string looped about alternate strands and tied to a heddle rod of sturdy arrowwood. The weft was wound around a bobbin shuttle of arrowwood and beaten in with a mesquite sword batten. If a narrow band was being woven, such as a belt, headband, or cradleband, the heddle was reduced to a loop of string that encircled alternate warps, and the weft

was driven home merely by pulling apart opposing sheds. The weaver, often a man, sat tailor-fashion before the loom. When the weaving approached the far end, he moved to that end and began anew, darning in the last few wefts with a slender stick.

The horizontal ground loom, while persisting today in northern Mexico, has disappeared from the American Southwest, driven out during the late nineteeth century largely by cheap factory-made cloth from eastern states. (The opening of the Southwest to trade similarly threatened Navajo weaving with extinction, but the ensuing tourist trade guaranteed the Navajos enough of a market to sustain a weaving interest [see Chapter 4].) The ground loom thrived best in wide-open spaces and where portability was at a premium—as with the Bedouin nomads. In such conditions the loom required and received no further "improvements."

THE VERTICAL LOOM
It is thought that the vertical two-bar loom originated in Syria or Mesopotamia as a method of stretching the crimpy and elastic wool warps for tapestry weaving. It seems probable that the discovery that wool readily accepted dyes inspired tapestry weaving, which in turn led to the development of the vertical loom as the most convenient tool for the purpose. It was easier, weavers learned, to pull the warps forward than up.

THE EGYPTIAN LOOM
Competition from Syrian tapestry weavers with their multicolored woolen yarns possibly inspired the development of the vertical loom during the New Kingdom in Egypt (c. 1570-1085 B.C.). The relatively late arrival of colored yarn—and hence tapestry weaving—in Egypt may have been due in part to the difficulty of dyeing linen, but there was also an ancient prejudice favoring white linen over wool as a purer, more acceptable garment for priestly and even secular uses. Wool was not unknown to the Egyptians, but since it was, as Apuleius said, "the excretion of a sluggish body taken from a sheep, [it] was deemed a profane attire even in the times of Orpheus and Pythagora; but flax, that cleanest production of the field, is rightly used for the most inner clothing of man." Various other explanations for wool's second-class status have been preferred, one of the more interesting of which involves the prevention of idolatry. According to Maimonides, the Israelites were forbidden to wear a mixture of linen and wool because, it was said, such attire was worn by heathen priests in hopes of a "lucky conjunction of the planets bringing down a blessing upon their sheep and flax."

Perhaps it was simply due to the nature of the fiber, but wool weavers in general were more experimental than linen weavers. For example, they are credited with the first application of twill weaving to cloth, previously associated only with mat making, in the Middle East—a development of no small importance to the weaving of wool. Aside from design considerations twills enabled the weaver to throw the shuttle from selvage to selvage, a difficult if not impossible task with wool plain weave. The setup for twill put more space—and thus less friction—between the kinky wool warps, allowing a shed large enough to send the shuttle all the way across.

The simplest loom-woven twill required three heddles, a development that takes us somewhat in advance of our story, as it is said to have developed in the Middle East during the third century A.D. (In Scandinavia 2/2 twills, probably developed on the warp-weighted loom, were being woven before the end of the Bronze Age. If there is a connection between Scandinavian twills and the 2/1 twills of the Middle East, probably developed on the two-bar loom, it remains unknown.) Since plain weave was an ideal weave for locking the slippery linen fibers together, technological innovation fell to the wool weavers with their stickier and more easily dyed yarns. Even with the introduction of silks into Egypt, the linen weavers saw no need to develop their loom beyond two harnesses. The

wool weavers, however, took the silk thread and experimented with it on their looms with their techniques for wool. The Egyptians were content to adapt only as circumstance made necessary.

Whatever the reasons for the delayed use of color in Egyptian weaving, whether religious, technological, or commercial, during the New Kingdom the Egyptians changed their weaving technique. The use of colored yarn coincided with the arrival of the vertical loom, and the two innovations were undoubtedly related. (Some experts consider the pressure of an increasing population a factor in the development of the vertical loom in Egypt. With less space available to the weaver for stretching his warp along the ground, the logical direction to go was up.)

The earliest examples of Egyptian tapestries were excavated from the tomb of Thothmes IV (1405 B.C.). The example in fig. 3-10 bears the name of his father, Amenhotep II, who ruled Upper and Lower Egypt from about 1449 to 1423 B.C. This particular fragment, a diaper pattern of alternating papyrus blossoms and lotus flowers, contains sixty warp threads per inch, is delicately rendered on a white ground in red, yellow, green, blue, and brown,

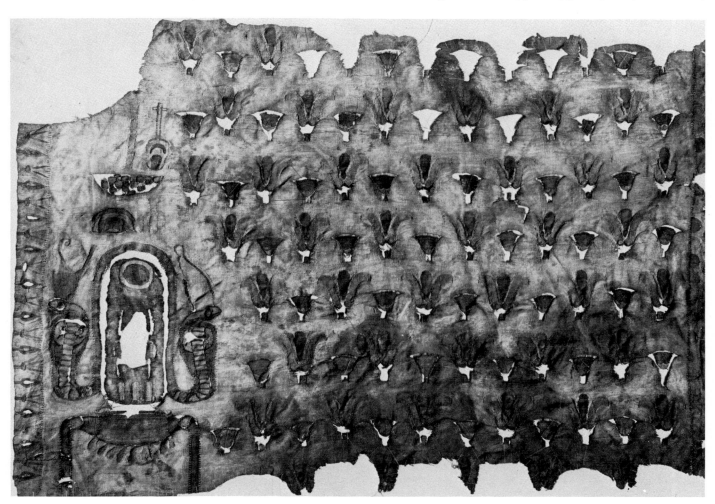

3-10: Tapestry of Amenhotep II from the Tomb of Thothmes IV, c. 1405 B.C., 11¼'' high x 16¾'' long. Egyptian Museum, Cairo.

white

white

3-11: Vertical looms from the Tomb of Thot-nefer at Thebes, XVIII Dynasty, c. 1425 B.C. Courtesy of Calderdale Museums Service.

and outlined in black. The loom on which it was woven probably resembled the wall painting in the tomb of Thot-nefer, a royal secretary of the 18th Dynasty, c. 1425 B.C. (fig. 3-11).

No reproductions of vertical looms from this period have been recovered intact, but it is possible to infer from the fragmentary evidence that both beams revolved (or perhaps the top beam could be lowered), that sheds were changed by means of a shed rod and rod heddle, and that

a long, heavy sword beater was used for beating down the weft. (That a weft fringe, characteristic of Egyptian ground-loom weaving, is not found in New Kingdom fabrics reinforces Bellinger's view that it was produced by uneven beating. With gravity providing the power behind the sword beater on the vertical loom, the weft received the same force from selvage to selvage.) Though various interpretations have been suggested, many of the details of the vertical loom, such as the disks on the uprights and the diagonal line crossing the upright to the left (fig. 3-11), remain unclear.

The Egyptians had perfected a loom that beat the weft _down_—a loom that could produce extraordinarily fine linens with thread counts as high as 280 x 80 per inch. Herodotus, commenting on this development, reminds us how unusual, at least in his experience, the Egyptian loom was: "The men sit at home at the loom; and here, while the rest of the world works the woof up the warp, the Egyptians work it down." Apparently Herodotus knew only the warp-weighted loom, but his observation also confirms what can be seen in fig. 3-11—that since the Middle Kingdom men had moved into the profession. Many of them undoubtedly were slaves working under the factory conditions described in the epigraph from the *Papyrus Anastasi* that introduced this chapter.

Weaving had become big business. One of the highest offices in the Pharaoh's administration was Director of the King's Flax. By 600 B.C. the Phoenicians were exporting Egyptian linen all over the Mediterranean and even as far north as Britain. It is believed that this commerce, well before the Roman occupation of Egypt in 30 B.C., carried the two-bar vertical loom to Rome sometime before the Christian era.

THE ROMAN LOOM

Weaving in Rome reflected her position as the greatest conglomerate of all time. At its zenith the Empire was bounded by Armenia, middle Mesopotamia, the Arabian Desert, the Red Sea, Nubia, the Sahara, the Moroccan Mountains, the Atlantic Ocean, the Irish Sea, Scotland, the North Sea, the Rhine, the Danube, the Black Sea, and the Caucasus. Among other things the extent of the Empire gave Rome access to the tapestries of Greece, the linens of Egypt, and the richly dyed woolens of the Near East. With such spoils available to her armies Rome felt no need for a lavish domestic production. With roads to build and conquests to be consolidated the weaver's art made but small progress at home.

Yet a textile industry flourished during the Empire period, with specialists for embroidery, fulling, felting, dyeing, and so on. Much of this industry, however, consisted of slaves working under factory conditions to provide clothing for the Roman troops. The guilds (their very presence indicated an active industry) controlled commercial production of imported linens and cottons, while woolens remained largely the province of household weavers. Household weaving received its official imprimatur when Augustus, reacting against the trend toward increasing richness in dress, set an example by wearing homespun garments woven by his sister, wife, and daughter. It is doubtful that Augustus' efforts did much to stem the growing tide of richness, but the tradition of household weaving, especially among the upper classes, did continue into the sixth century A.D. Local industry was generally limited to the production of everyday clothes, and in the later years of the Empire even peasants were buying their clothes ready-made.

With the expansion of the Empire Rome grew increasingly parasitic, living more and more off the wealth of conquered lands while producing less and less at home. Her taste for sumptuousness was indulged with silks and colorful tapestries in wool and gold carried over the caravan routes from the East. By the time the silks reached Rome after a hazardous journey that often took three years or longer, they were literally worth their weight in gold. (The secret of silk production did not reach Byzantium until the sixth century.) Yet silks were not used sparingly. The luxurious Tyrian purple was so desired that the murex, the shellfish from which the dye was extracted, was threatened with extinction. Consequently, a law was passed permitting only the nobility to wear what came to be called "the royal purple."

Among the excesses of the Empire was a tendency to supplant beauty with richness, and it was to this end that embroidery began to supplant tapestry in Roman times. Asterius, bishop of Amasia, described the absurdity to which this was carried during the fourth century: "When men appear in the street thus dressed, the passers-by look at them as at painted walls. Their clothes are pictures, which the little children point out to each other. Here are lions, panthers, and bears; there, rocks, woods and huntsmen. The most saintly wear likenesses of Christ, his disciples, and his miracles. Here we see the marriage of Galilee, and the pots of wine; there, the paralytic carrying his bed, the sinner at the feet of Jesus, or Lazarus raised from the dead." Although much of this finery was probably embroidered, tapestry too had become the equivalent of frescoes on cloth.

The looms on which Roman tapestries were "painted" are thought to resemble those in Ovid's description of the weaving contest between Athena and Arachne cited in the previous chapter. Although it is not absolutely certain, it is generally believed that Ovid was describing a two-bar vertical loom. His detail is regrettably less generous regarding the looms than the design of the tapestries woven on them.

Four representations of Roman looms have come down to us, two of which (figs. 3-12 and 3-13) are reproduced here. Again, the detail leaves much to be desired. Although it is not evident here, the width of the loom often must have been greater than its height, as some fabrics required a web some six to eight feet wide. The top beam of the Roman loom probably did not revolve but could be lowered through slots in the uprights as the weaving progressed and was wound on the lower beam. A heddle rod supported by notched pegs in the uprights opened one shed; the second was presumably opened by a flat shed stick turned on edge. (Margrethe Hald, writing about ancient Danish textiles, interprets this rod not as a heddle rod but as a transverse cord, or rod, around which the warp ends are wound to form the joint of tubular weaving, a variant technique on the two-bar loom [see fig. 3-16].) The repre-

3-12: Two-beam vertical loom from the Forum of Nerva, Rome. Deutsches Archäologisches Institut, Rome.

3-13: Wall painting of a two-beam loom from the *Hypogeum* of Aurelii, Rome, possibly representing Penelope at her loom. Fototeca Unione, Rome.

sentations indicate that the uprights stood on wooden blocks, or feet, and that the weaver sat at her work. She used the shed rod and heddle for the ground weave while working the designs in freely by hand.

THE PALESTINIAN LOOM

A similar loom was used in Palestine during Talmudic times (C. A.D. 500–1100), though it was said that women should not weave on it for fear of appearing immodest by exposing an armpit. (The horizontal loom may also have been in use at that time.) Weaving in Palestine began as a domestic occupation, but as it became commercial, guilds were formed and, as was common elsewhere when some hope of profit appeared, men took over. The profession was considered somewhat disreputable, for the weavers lived in the worst section of Jerusalem and, according to R.J. Forbes, were regarded as "rough, treacherous and dishonest as they had frequent contact with women."

Nonetheless, the profession and the vertical loom, though no longer as common as the treadle or ground loom, have survived into the twentieth century. The modern Palestinian vertical loom (also found in Syria) consists, like the Roman loom, of two uprights with a fixed bottom beam and a top beam that can be raised or lowered by slots in the uprights. It differs from the Roman loom, however, in several important respects, the most interesting of which is a *third* beam (fig. 3-14) behind and away from the loom, attached to a wall or other support, that affords the weaver a longer warp and a unique means of adjusting the tension. While it employs the traditional shed rod and heddles, the heddle rod is fastened to its supports and functions like the fixed heddle arrangement of the Bedouin ground loom but vertically. A third distinctive feature is the way in which the loom is warped (fig. 3-15). With the ball of yarn lying on the ground, the warp is wound alternately in continuous loops around the beams and back over a warping rod that is fastened by one end to an upright. Thus, with the weaver sitting in front, the entire warp can be slowly shifted around as the weaving progresses. The beating in is done with a heavy comb and both a sword and a pin beater. Upon completion of the fabric the warping rod is removed, leaving the cloth with a looped fringe at either end.

3-14: Modern Palestinian three-beam loom (only two of which are visible here). Courtesy of Miss Elisabeth Crowfoot and the Palestine Exploration Fund, London.

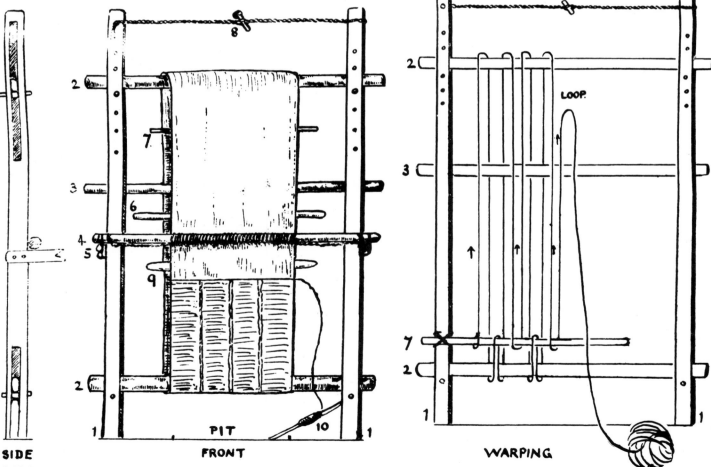

SIDE   FRONT   WARPING

3-15: Diagram of Palestinian three-beam loom and warping technique. (1) Uprights. (2) Upper and lower beams. (3) Third beam. (4) Rod Heddle. (5) Support for heddle rod. (6) Shed rod. (7) Warping rod. (8) Peg twisting cord above loom. (9) Sword beater. (10) Spool or bobbin. Courtesy of Miss Elisabeth Crowfoot and the Palestine Exploration Fund, London.

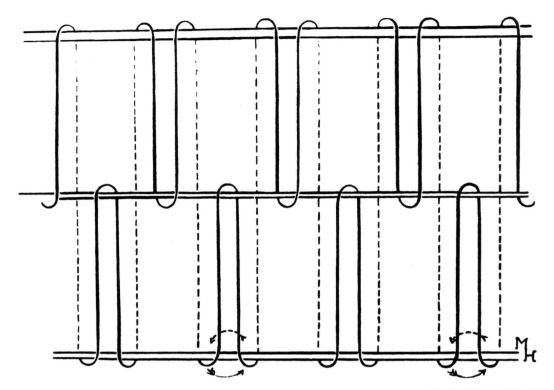

3-16: Diagram of tubular warping technique on early Scandinavian textiles. Courtesy of Margrethe Hald and the Danish National Museum, Copenhagen.

3-17: Tubular-woven skirt from Huldremose showing locking of looped warp ends around transverse cord. Photograph courtesy of Danish National Museum, Copenhagen.

Analogous warping techniques have turned up in places as far distant as Tibet, Scandinavia, South America, and the Northwest Coast of North America. Such a curious correspondence—whether related or merely parallel—reflects one of the most persistent problems of this subject: that of independent invention. Was it possible that such a specialized technique was invented separately in places so far removed from one another? Or were the seeds of influence carried across now-obscured trade and migration routes? Experts have yet to agree on the answer. Based on textile finds, one of these experts, Margrethe Hald, has dated Scandinavian examples of the tubular warping technique (fig. 3-16) to the Celtic Iron Age, c. 500 B.C. to the early Christian era. She finds it unlikely that places so far apart should have invented the same technique. What is more probable, Hald suggests, is that tubular weaving was a developmental stage in the evolution of the craft.

The advantage of tubular warping was that the weaver could double (or in the case of the Palestinian loom even triple) the length of the cloth without increasing the height of the loom. Further, since the warp could be shifted in either direction, one could weave from the bottom up, the top down, or, presumably, even horizontally. Nor was it necessary to remove the transverse juncture cord upon completion. If a tubular fabric was desired, the cord was simply left in (fig. 3-17).

3-18: *Clal-Lum Women Weaving a Blanket.* Oil painting by Paul Kane (1810–1871). Courtesy of the Royal Ontario Museum, Toronto.

THE SALISH LOOM

Just south of the Chilkat area in North America, along the coast of the state of Washington, the Salish Indians were weaving on a loom (fig. 3-18) virtually identical to the Scandinavian loom described by Hald. (The Salish also used a loom similar to that of the Chilkat tribe on which the weft was twined across the warp threads.) The typical Salish loom consisted of two planks sunk in the ground about six feet apart. Each plank, often scalloped along the top, had two slots to accommodate the upper and lower beams. The beams fit loosely in the slots, and tightening the warp was merely a matter of driving wedges into the slots to force the bars farther apart.

Instead of weaving with sheep's wool, as did the Scandinavians, the Salish used goat's wool and the hair from a small Pomeranian-type dog. (Apparently the dog resisted the haircut, for the Salish swung him in the air before shearing to make him dizzy.) These fibers were often spun together with duck or goose down on large (36″ long with an 8″-diameter whorl) spindles. Most of their blankets were 2/1 twills woven with the fingers without the aid of heddles or bobbins (fig. 3-19). The only tool was a sword beater for opening the shed and beating in the weft.

Some experts believe that the Salish loom is culturally related, as the most northern example, to a loom used in the Antilles and South America. The Waiwai and Guayos tribes of British Guiana, for example, wove loincloths up to six feet long and five to six inches wide on a similar loom made of cane that was lashed together into a frame (fig. 3-20). The weaver leaned the loom against a house or tree and braced the uprights between his toes as he wove. (Compare this to the Arawak loom described in Chapter 5.) Ronald L. Olson, writing in *American Anthropologist* in 1929, traced the somewhat random distribution of this type of loom northward through the southeast United States, the Plains area, and the Columbia basin to the Salish on the coast (fig. 3-21).

Perhaps one relative, more akin to the vertical Salish loom than other horizontal looms because of the tubular warping, is a horizontal loom still used in northern Mexico for weaving skirts, wraparounds, serapes, and wide blankets (fig. 3-22). The loom bars here are fastened to four uprights pounded into the earth. On the Tarahumara loom in fig. 3-23 the beam is supported by two side rails with forked branches that both elevate the beam and hold it secure. The breast beam is lashed to the rails alongside the weaver. Note the long stick bobbins and sword beater leaning against the rail. The backstrap loom (see Chapter 5) was also used for tubular weaving.

3-19: Salish robe woven of dog and mountain-goat wool, collected in the 1840s. The central (white) portion is woven in the twilled-checker technique, and the sides are woven by plain twining, with wefts slanted to form diagonal lines. National Museum of Natural History, Smithsonian Institution.

3-20: *Left:* Loom for weaving seamless garments. Guayos Indians, British Guiana. The selvage threads carried out to the side of the loom frame are woven with their own set of heddles to strengthen the selvage. Courtesy of the American Museum of Natural History. *Right:* Diagram of Waiwai loom for weaving tubular cloth, showing the unusual method of double weaving the selvages. Courtesy of the Smithsonian Institution.

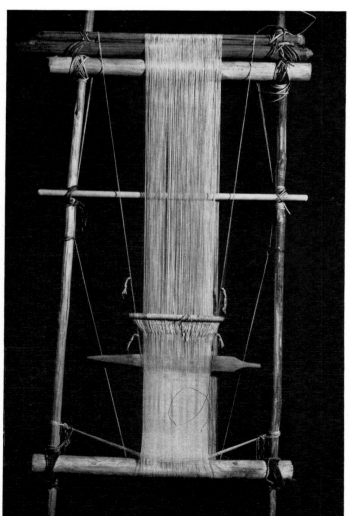

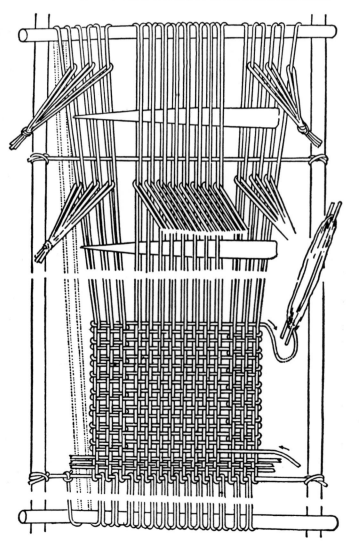

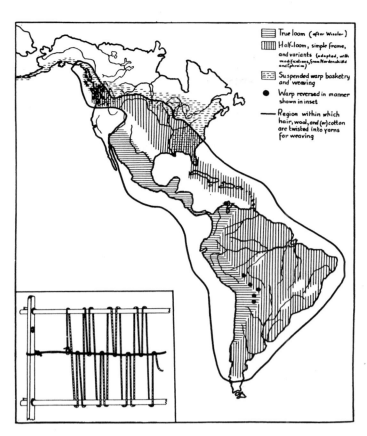

3-21: Tentative areas of various weaving techniques in the Americas. Inset: Diagrammatic representation of the method of winding the warp so that its direction is reversed. Reproduced by permission of the American Anthropological Assocation from *The American Anthropologist*, 31 (1), 1929.

3-22: A heavy blanket being woven by a Mayo Indian of Huatabampo, Sonora, Mexico, 1938. The loom, which produces a tubular fabric, is not common in Mexico. Photograph by Donald Cordry.

3-23: Tubular weaving by Tarahumara Indian in Wachochi, Chihuahua, Mexico, 1953. Photograph by Irmgard W. Johnson.

son tan brabos y jus ti cieroy mal pratan los yns y haze tra uajar co un palo estes kirey no en las do rinas noay ue medio

3-24: Peruvian vertical loom (1613) showing the use of a comb for beating in. From the MS of the Peruvian Indian Felipe Hnaman Poma de Ayala, *El primero i nueva coronica i buen gobierno*. Courtesy of the American Museum of Natural History.

THE SUB-SAHARAN AFRICAN LOOM

It would be superfluous to detail every country or people that either adopted or adapted the two-bar vertical loom for its own use. (See the Araucanian loom in Chapter 5 for another example.) It was widespread and ranged from a Peruvian tapestry loom (fig. 3-24) to a primitive mat loom in Central Africa to the husky rug looms of India and Persia to the high-warp tapestry looms of Renaissance Europe. Any attempt to trace the genealogy of the two-bar loom to its innumerable habitats would inevitably bog down in a scrutiny of textiles far beyond the scope of this book. And even then there would be much room for disagreement, because only in certain cases can one divine on which type of loom a particular textile was woven. Yet the looms of some of these areas are worth examining either by virtue of their unusual and interesting features or of their influence on the evolutionary history of the loom. It is with the former

in mind that we consider briefly the vertical loom of sub-Saharan Africa.

Its origins are obscure. Some say that it might have been introduced by Portuguese traders. Others believe that a more sophisticated loom of Asian origin traveled a route through Arabia and south into Africa. If the loom was more sophisticated at the start, by the time that it had penetrated deep into the continent, it had degenerated into a more rudimentary form. It seems certain that textiles were produced, probably on looms, well in advance of European contact.

Most of the evidence for the vertical loom in sub-Saharan Africa comes from West and Central Africa. Where the more sophisticated toe-treadle—or strip—loom is known (see Chapter 6), the vertical loom is used exclusively by women; but where the strip loom has not been introduced, as in the Congo River basin in Zaire, men still weave raffia cloth and mats, usually without selvages, on a vertical two-bar loom. (Mary Elizabeth King believes that the loom used exclusively by women is the earlier of the two.) It resembles DuChaillu's 1867 illustration depicting an Ishogo weaver at his loom, enjoying a well-traveled smoke while he works (fig. 3-25). The loom has no uprights. The warp beam is suspended from the ceiling, and the tension is provided by stakes in the ground that secure the cloth beam. What appear to be two heddles are in fact one (fig. 3-26), and the needle shuttle that the weaver holds in his right hand serves as the shed stick and sword beater as well.

In another version of this loom (fig. 3-27) the warp is stretched at about a sixty-degree angle to the ground, with the weaver sitting inside the angle under the warp. The warp beam is lashed by spiral loops to a bar supported by two uprights. The warp threads are tied in bunches to pairs of cords, which are in turn fastened to the warp beam. On the Babunda or Bapindi loom from the same area these fastenings are tied in the curious manner illustrated in fig. 3-28. All the upper ends of the cords are carried along the warp beam to the last pair of cords on the right, where they are knotted together. If the Babunda system was used to secure the warp threads to the cloth beam, it probably resembled that shown in fig. 3-29.

The heddle, unlike the Ishogo weaver's, consists of a single piece of split cane (fig. 3-30), while the shed bar and beater-bobbin are fashioned out of palm ribs. It appears that the bobbin, notched at one end, is first inserted through the shed, hooks one of the weft threads draped over the line to the left of the weaving, and draws it back through the shed. The bobbin is then reinserted into the shed, where it performs its auxiliary service as beater.

Besides the raffia loom a vertical cotton loom has been developed in Africa. It is often referred to as the woman's loom since it is usually found where men weave on the horizontal strip loom. One example (fig. 3-31) shows the

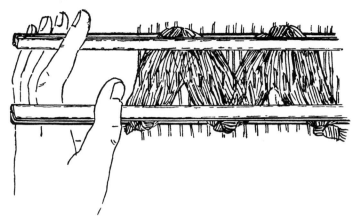

3-26: Method of holding the heddle on Ishogo vertical loom. Courtesy of Calderdale Museums Service.

3-25: Ishogo man weaving on vertical mat loom. From DuChaillu, *Ashango Land*, 1867.

3-27: Man's vertical raffia loom, Kuba, Republic of Zaire. Photograph by Barbara W. Merriam.

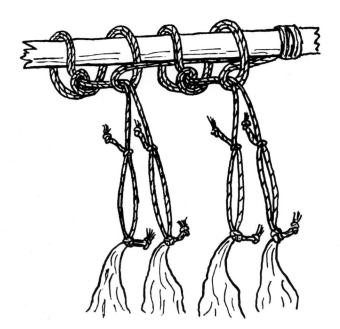

3-28: Method of attaching warp to the warp beam on the Babunda loom. After T. A. Joyce, "Babunda Weaving," *Ipek*, 1925.

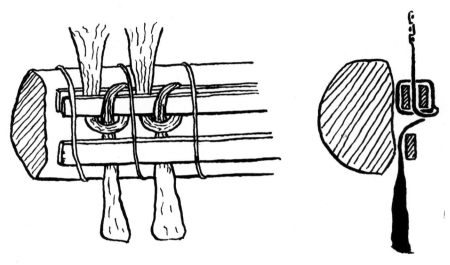

3-29: Babunda method of securing warp to the cloth beam. After T. A. Joyce, "Babunda Weaving," *Ipek*, 1925.

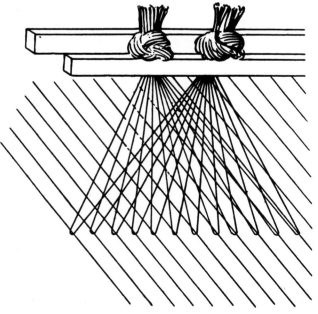

3-30: Babunda heddle arrangement. Schematic drawing omitting warps not contained in heddle leashes. After T. A. Joyce, "Babunda Weaving," *Ipek*, 1925.

3-31: (a) Ibo woman's loom from Akwete, Nigeria. Denver Museum of Natural History. (b) Details of an Akwete loom. (1) Stick used to hold warps in position. (2) Temple, a grooved stick to keep the weaving width even as weaving progresses. (3) Shed stick. (4) Six string-loop heddles, which control warps for the twill pattern. (5) String-loop heddle controlling even warps for plain weave. (6) Supplementary shed stick helping to hold odd warps forward when needed. (7) Batten. (8) Two stick bobbins on which plain-weave and brocade wefts are wrapped, c. 4' long. Museum of International Folk Art.

uprights inserted into holes cut in the top and bottom bars, but other illustrations show the bars lashed to the uprights. In the Ibo loom shown here a second bar is used as the warp beam, but just as often the warp will be wound directly over the top beam instead of using a second bar. It is wound in a continuous spiral around the beams so that the entire warp can be rotated as the weaving progresses. The finished cloth, as on the Salish loom, measures twice the distance between the beams. Cloth from the women's looms is traditionally used just as it comes from the loom. Alternatively, it may be sewn together with similar pieces to make a larger fabric for blankets, mantles, or wraparounds; it is not tailored.

THE LOOM IN THE DARK AGES
As successive Barbarian invasions swept over the disintegrating Roman Empire, commerce and industry crumbled and weaving reverted to a household occupation. Records concerning weaving and looms are sparse during the en-

3-32: Vertical loom shown in the Utrecht Psalter, 9th c. A.D. After E. T. Dewald, *The Illustrations of the Utrecht Psalter*, 1932. Courtesy of Princeton University Press.

suing Dark Ages in Europe, but evidence of the two-bar vertical loom crops up in several places.

The Utrecht Psalter of the ninth century contains a poorly defined illustration of the vertical loom as it must have appeared in the early Middle Ages (fig. 3-32). Very little can be distinguished except that the uprights are supported by a crossbeam that is fastened to the canopy supports. The weaver, kneeling or sitting before the loom, holds a comb—with the teeth perpendicular to the handle—for beating down the weft. Her assistant, standing to the right, holds a pair of scissors. At the left a new warp is being prepared.

Further north in Oseberg, Norway a ninth-century loom has been unearthed from the remains of a Viking ship (fig. 3-33). It is a small loom with a base that probably sat in front of the weaver on a table. The indented section of the lower beam—presumably the width of the warp—is only 33–34 cm. Neither the upper nor the lower beam revolved. The tension was maintained by adjusting the lower beam via the holes in the lower part of the uprights. It is possible, though no proof exists, that the loom accommodated a

tubular warp. It was probably not used for making everyday cloth (it was too small) but was a specialized tool for figured fabrics.

It is unclear how popular the vertical loom actually was during this time, but in 1070 Theophylact, a bishop of Bulgaria, assumed it was common throughout southern Europe: "Others say that in Palestine, they weave their fabrics not as with us, having warp threads above and weaving below with the bobbin and thus mounting; but on the contrary, the warp threads are below and the web is woven from above." This statement is often quoted to argue the existence of the two-bar vertical loom in Europe during the eleventh century, but it is not certain from Theophylact's phrase "not as with us" how widespread in Europe the "us" was—for it was the eleventh century that saw the introduction of the horizontal treadle loom into Europe (see Chapters 6 and 8).

One final illustration of the vertical loom's tenure in Europe comes from the manuscript *De universo* by the German theologian Hrabanus Maurus, also dating from the eleventh century (fig. 3-34). The position of the weaver

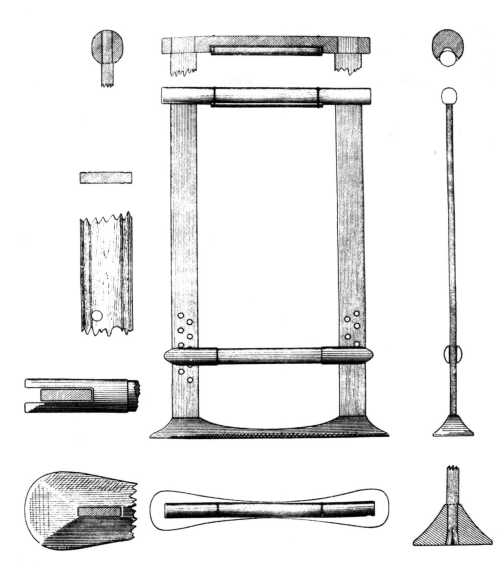

3-33: Diagram of the Oseberg loom, 9th c. A.D. After Oseberg II.
© Universitetets Oldsaksamling, Oslo, 1928.

3-34: Loom shown in the Hrabanus Maurus MS *De Universo*, 11th c.
A.D. Courtesy of Bildarchiv Preussischer Kulturbesitz, Berlin.

suggests that he is sitting with his legs in a pit, though one must be careful of interpreting these illustrations too literally. Again, neither beam can revolve, as each is fitted around the uprights by crotches and presumably pegged in place. (Hoffmann believes that the many dots on the uprights represent holes for adjusting the height of the beams.) Though no indication of shed rod or heddle is shown, the wavy line above the weaver's head is probably a spacing cord to keep the warps in order.

Thus, what sparse evidence there is suggests that the two-bar vertical loom remained virtually unchanged from Roman times until it was supplanted by the horizontal loom during the eleventh and twelfth centuries. (No evidence has been found to show that the vertical loom was used for weaving ordinary cloth in the Middle Ages. Although this does not mean that it wasn't done, the illustrations from the period all show the horizontal loom being used for that purpose.) The one notable exception was the tapestry loom, which developed into a highly specialized tool during the late Middle Ages and is still used today.

THE TAPESTRY LOOM

European tapestry weaving did not originate in the Middle Ages when the horizontal loom took over the production of ordinary cloth. We have already seen poor Penelope weaving and raveling a tapestry to deceive the suitors until the return of Odysseus. In the *Iliad* Helen is found weaving at her loom her own tragic story: "The Trojan wars she weaved, herself the prize, / And the dire triumph of her fatal eyes." And in the *Metamorphoses* Athena and Arachne engage in a weaving contest that demands the utmost in tapestry skill.

Following the Greek and Roman periods, however, tapestry weaving went into hibernation for about a thousand years, during which time the art of large wall pictures fell to painters and embroiderers. What little tapestry weaving continued was practiced either in households or in monasteries and convents, where it was safe from the ravages of war. Nuns probably passed the art along to noble ladies who, not unlike Penelope, sat weaving away the hours until their lords returned from the wars. It is not surprising that tapestry found a home in the church, for in the twelfth century the church owned half the land in England and even more on the continent. But the age of chivalry and romance had begun to sweep across Europe, and the influence of the church was on the decline. Perhaps it was then, in the twelfth century, that tapestry weaving broke out of the church and into the craft guilds.

Although a great deal has been written about the importance of the medieval and Renaissance cloth trade, very little note has been taken of its looms. It was not until the eighteenth century when Diderot conceived his illustrated encyclopedia that much attention was paid to the tools of the trade.

The vertical tapestry loom (fig. 3-35)—what is known as the *haute lisse*, or high-warp, loom—is first mentioned in a Paris ordinance of 1302 regulating tapestry weavers in a manner similar to other weavers. The *haute lisse* loom probably existed somewhat earlier, perhaps as early as the ninth century, but it didn't come into its own until the Gothic period and the full flowering of pictorial tapestries. (The counterpart of the *haute lisse* loom was a horizontal treadle loom called the *basse lisse* [low-warp] loom. Because feet were employed as well as hands, the weaving went faster, but the *basse lisse* loom had certain built-in disadvantages: its very speed was conducive to error, and the only way to see the front of the finished fabric was by looking through the warp to a small mirror placed underneath. In spite of these disadvantages the *basse lisse* loom gained general acceptance by the early sixteenth century. Even so and despite the fact that products of this loom were almost impossible to distinguish from *haute lisse* tapestries, the *basse lisse* loom was never accorded the respect of its older, more venerable companion.)

The weaver at the *haute lisse* loom sat behind the warp with the *lisses* (leashes) suspended overhead from a rod or rods (fig. 3-36). The design was traced on the warps with chalk (fig. 3-37) from a pattern that the weaver hung nearby for ready reference. The wefts were wound around small pointed bobbins that were also used, along with a comb, for pressing in the inserted threads. (The *basse lisse* loom was equipped with blunt, not pointed, bobbins, and the weft was beaten in with the comb alone.) Weaving in this way was slow work. On the *haute lisse* tapestry loom a weaver was happy to complete three yards in a year. The loom, as illustrated by Diderot (fig. 3-38), had beams that rolled up the finished tapestry as it unrolled additional warp, so the weaver never saw the total effect of his work until it was completed.

The Golden Age of tapestry is commonly said to have been the transition period between the Gothic and the Renaissance periods, c. 1450–1550. After that painters such as Raphael were employed to design tapestries, and it became the vogue for tapestries to imitate paintings. By the eighteenth century, when the painter Oudry introduced the idea of blending colors in tapestries to approximate paintings even further, the triumph of the painters was complete. The Flemish weavers of the Golden Age created their masterpieces with only twenty to thirty different colors; the Gobelin workshops today use over sixteen thousand different shades for *haute lisse* weaving alone.

Although this is a book on looms, not textiles, a few words should be said about the Gobelins—especially since in some quarters the term is regarded as almost synonymous with tapestry. The Gobelin family, in spite of the associations connected with the name, never produced a single tapestry. Jehan Gobelin, originally from Rheims, was what was known as a "merchant dyer of scarlet" who

3-35: Back and front of a small model tapestry loom (haute lisse).
Victoria and Albert Museum, London.

3-36: Haute-lisse tapestry loom. After Diderot, *L' Encyclopédie, Recueil de Planches*, Vol. IX, ''Tapisserie de Haute Lisse des Gobelins,'' Pl. IX.

3-37: Design traced with chalk on warps of tapestry loom. After Diderot, *L'Encyclopédie, Recueil de Planches,* Vol. IX, ''Tapisserie de Haute Lisse des Gobelins,'' Pl. XI.

3-38: Tapestry looms in the Gobelin workshop. After Diderot, *L'Encyclopédie, Recueil de Planches,* Vol. IX, "Tapisserie de Haute Lisse des Gobelins," Pl. I. The Beinecke Rare Book and Manuscript Library, Yale University.

in the mid-fifteenth century set up shop on the banks of the Bièvre in a suburb of Paris. Over the generations the Gobelin dyers became so successful that by the seventeenth century they felt that dyeing was beneath their dignity, and they turned instead to the world of finance. It was two Flemish weavers, Marc Coomans and François de la Planche, who established tapestry weaving chez Gobelin, first near the shop in 1601 and later in 1630 in the shop itself. The name of the workshop stuck, and in 1662 when Colbert, the minister of Louis XIV, united the Paris workshops under one head—*Manufacture Royale des Meubles de la Couronne*—the general upholstery workshop was popularly called the *Manufacture des Gobelins.* The shop was closed in 1694 and reopened in 1697, dedicated to the production of fine tapestries. The name "gobelin" became so firmly associated with tapestry weaving that it became a general term for any picture tapestry—including one woven in Flanders even before Jehan Gobelin set up shop on the banks of the Bièvre.

From ancient Egypt to seventeenth-century France is a long distance to travel in one chapter—especially with side excursions to the Near East, the American North- and Southwest, and Africa. During that journey of some seven thousand years the warp of the two-bar loom has been stretched out along the ground, tilted up at various angles, and hoisted upright. The weaver has sat on the cloth as she wove it, leaned over it, and even sat just about under it. The warp during this period has been fastened to the two beams, wound endlessly around them, looped over them and around a transverse cord or rod, tied to heading cords, and stretched over an auxiliary third beam. But in all these variations, from Egypt to Peru to North America to Scandinavia to Palestine and Rome, the shed was always formed with a shed rod and heddle. The two-bar loom and the shed rod and heddle: it was a simple, effective, and enduring combination. And before leaving it for the revolutionary advance of a shed changed by footpower, we have a few well-known looms left to consider—first those of the American Southwest.

4. Pueblo and Navajo Looms

Oh, our Mother, the Earth; oh, our Father, the Sky,
Your children are we, and with tired backs
We bring you the gifts that you love.
Then weave for us a garment of brightness;
May the warp be the white light of morning,
May the weft be the red light of evening,
May the fringes be the falling rain,
May the border be the standing rainbow.
Thus weave for us a garment of brightness
That we may walk fittingly where birds sing,
That we may walk fittingly where grass is green,
Oh, our Mother, the Earth; oh, our Father, the Sky!
—Tewa Indian prayer

Navajo weaving may be an anachronism in modern America. It may be unprofitable; it may be exploited; it may be debased through imitation. Ultimately it may vanish altogether. Yet the ordinary Navajo loom, essentially unmodified since pre-Columbian days, represents the culmination of native loom development on the American continents. (Its rival for that honor, the backstrap loom, also used by the Pueblos and Navajos, will be considered separately in the next chapter.) While the Navajos have accepted many products of western technology, such as the automobile, they have steadfastly refused to "improve" (i.e., mechanize) their loom. They have not found it artistically necessary. In fact, the mechanized loom would probably prove fatal to Navajo weaving—as almost happened with the introduction of Germantown yarns in the late nineteenth century. Both Navajo and Pueblo looms are two-bar looms similar to some of those discussed in Chapter 3.

ORIGINS

The Navajos attribute the origin of their weaving to Spider Woman "in the beginning." Not only was Spider Woman responsible for introducing weaving to the Navajos, but she continued her influence through the ritualistic preparation of female babies for a future at the loom. For according to legend Spider Man had said: "Now you know all that I have named for you. It is yours to work with and to use

following your own wishes. But from now on when a baby girl is born to your tribe you shall go and find a spider web which is woven at the mouth of some hole; you must take it and rub it on the baby's hand and arm. Thus, when she grows up she will weave, and her fingers and arms will not tire from the weaving."

Without disputing myth or legend, most historians state that the Navajos learned to weave from the Pueblo Indians at the end of the seventeenth century. They borrowed the Pueblo loom and the Pueblo techniques and even acquired their wool by raiding Pueblo flocks. The Pueblo people resided on the plateau consisting of northern New Mexico and Arizona and southern Colorado and Utah (fig. 4-1). They were named by the *conquistadores* who discovered them living in their high-rise villages, or *pueblos* (fig. 4-2), in 1540 during the Spanish northern quest for the fabled seven golden cities of Cibola. But the Pueblos did not "invent" the loom—though they might have been the first to add heddles to a vertical frame. The American loom has a history that, like the European, dates into the unknowable—or at least presently unknown—past.

While the true loom (i.e., a loom with heddles) in the Southwest is generally associated with the introduction of cultivated cotton into the area c. 700 A.D., Charles Amsden believes that other antecedents made the appearance of the true loom possible. In *Navajo Weaving: Its Technic and Its History* he traces its origins to the "supporting stake" of the Basketmaker II Period (roughly from the beginning of agriculture to A.D. 400), on which finely twined bags of yucca fiber and apocynum (Indian hemp) were "woven" (fig. 4-3). (A similar method was used by the Osage Indians for small bags and by the Northwest Coast Indians for constructing baskets.)

To weave such bags, six or more warp strings were bunched in the center so that the ends radiated outward like the spokes of a wheel. This warp foundation was hung upside down to keep the strings from tangling, and the weft was intertwined circularly, working outward (downward in this case) from the bottom. To make the bag larger at the center, more warp strings were added, and to narrow it at

4-1: Map of Pueblo area in the Southwest. Copyright © 1969 by Virginia More Roediger, reprinted by permission of the University of California Press.

the neck, the number was reduced. Two weft strands were crisscrossed between the warps in the twining technique. The result was a perfectly round, seamless bag.

The step from this single point of suspension to a multiple-point suspension, as typified by the Chilkat warp-weighted loom, was a small one. On the single-beamed loom the warps dangled freely in preparation for twining, plaiting, or twilling. Whether this particular loom made its way to the Southwest or evolved there from the Basketmaker culture is not known, but the Paiutes of southwestern Utah and northwestern Arizona had a one-beamed loom on which they twined rabbit-skin warps into blankets (fig. 4-4).

The next stage, a fixed warp frame, discussed earlier as the Salish loom, consisted of two parallel beams supported by uprights. The warp was wrapped in a circular fashion around the beams and held firmly in place for the fingers to manipulate. With the change from a free to a rigid warp the direction of weaving also changed. Instead of being pressed up the weft was now beaten down.

From this point the development becomes even more theoretical. Some believe that the impulse to heddles derived from the Peruvian backstrap loom from early Tiahuanacan times (see fig. 5-4), which was equipped with a long, slender stick for darning the weft into the warp. But the evolutionary lines are blurred by apparent developments in other parts of the Americas. Textile impressions on pottery indicate that a great many fabric techniques were developing in the New World, with weaving sparsely represented among them. Pottery fragments from Utah,

4-2: Multistoried dwellings at Zuni pueblo, c. 1899. History Division, Los Angeles County Museum of Natural History.

4-3: Attu weaver working upward on suspended warp. From *Indian Basketry* by O. T. Mason. Reprinted by permission of Doubleday & Company, Inc.

4-4: Indian twining a rabbit-skin blanket. Drawing by Zia artist Velino Herrera (Ma-pi-wi).

Ohio, New York, and Tennessee suggest that some kind of twining frame was being used before the arrival of cotton in the Southwest. A true loom with heddles probably did not exist outside the Southwest and the Northwest coast until European contact.

It is probable that cotton, the spindle, and the loom with heddles arrived in the Southwest together as part of the same technological complex. Most authorities believe that cotton made its way north to Arizona and northern Mexico from the Mayan centers long before the Aztec period, but they are uncertain as to its precise origin. The confusion results from the varieties of cotton that have been found: one kind, *Gossypium barbadense*, in Peru; another kind, *G. hirsutum*, in southern Mexico and Guatemala; and a third kind, *G. hopi*, in the Southwest. The latter two seem to be related. While strands of cotton have been found that might date back to the time of Christ, no loom-woven cloth has turned up that can be dated earlier than the Pueblo I Period, A.D. 700–900.

Cotton provided the finest but by no means the only fiber for weaving. The Pueblos also continued to spin the yucca fiber for twining bags, belts, sandal cords, and foundation cords for fur and feather string blankets. Apocynum (Indian hemp), a more pliable material from the milkweed family (sometimes confused with softer and more finely shredded yucca), was used for sandal wefts, twined bags, blankets, and aprons. But until the Spanish pushed north with their sheep in 1540, cotton remained the staple fiber that it had been since its introduction 800 years earlier. (Even the first sheep were not used for wool but for mutton. The shearing of sheep to obtain yarn did not occur until about 1600.)

Wool, when it did arrive, almost completely displaced cotton as the staple fiber. Cotton fields were gradually converted for sheep grazing until only the Hopi continued to cultivate what came to be called *Gossypium hopi,* and they grew it mainly for ceremonial clothing. But by that time the Classic or Golden Age of the Pueblo Indians (Pueblo III, A.D. 1050–1300) was behind them, and they were on the decline. The turning point was a disastrous twenty-two-year drought that began in 1276. The drought, together with increased pressure from their Athapascan neighbors to the north (Apaches and Navajos), forced the Pueblos to migrate from their northern population centers, and a cultural slump ensued from which they have never recovered. The arrival of the Spanish in 1540 only exacerbated their problems.

THE PUEBLO VERTICAL LOOM

The kind of loom that originally accompanied cotton into the Southwest was most probably the backstrap loom, generally regarded as the oldest heddle loom in the New World. (Though it might well have been the horizontal ground loom that was apparently used in northern Mexico and later by the Indians of southern Arizona.) In the Americas backstrap-loom weaving is usually associated with women, probably originating in Peru. Yet among the Pueblos it was the men who wove. Perhaps this developed from the southwestern tradition that men twined rabbit-skin blankets on the free-warp loom. Since men were responsible for hunting and skinning the rabbits, the step from preparing the skins to twining them into blankets is not too unusual. Or, again, perhaps it was an instance of men taking over a craft following the introduction of a new tool.

Women, however, helped out with the spinning and ginning of cotton. Although it varied from village to village, the most popular method of ginning cotton was to flail it between two blankets with three sticks tied together at one end. The fibers stuck to the blankets and the seeds fell free. Another method was to pick the seeds out by hand and then to beat the cotton on sand until it was clean and fluffy. Among the Maricopa and Papago Indians the cotton was ginned with a bow (see fig. 5-41), a nonnative tool of Asiatic origin that was probably introduced by Spanish or Portuguese priests who were familiar with its use in parts of

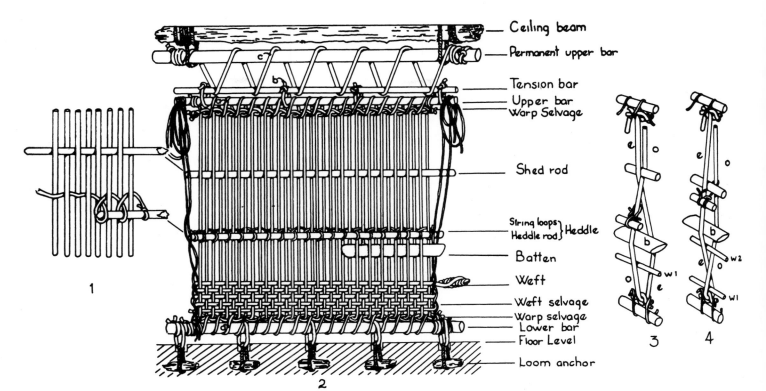

Ceiling beam
Permanent upper bar
Tension bar
Upper bar
Warp Selvage

Shed rod

String loops
Heddle rod } Heddle
Batten
Weft
Weft selvage
Warp selvage
Lower bar
Floor Level
Loom anchor

1

2

3   4

4-5: Schematic drawings of the vertical loom, rigged for plain weave. (1) Rigging of string loops to the heddle rod. (2) Loom and its working parts. (3, 4) Functions of shed rod and heddle in changing sheds. In (3) the heddle is shown pulled forward, and the batten (b) inserted behind (o) and turned sideways to open the shed for the first weft (w 1). In (4) the shed rod is shown forced down against the heddle loops, and the batten holding open the shed thus formed. The second weft (w 2) is in position. Reproduced from Kate P. Kent, "The Cultivation and Weaving of Cotton in the Prehistoric Southwestern United States," *American Philosophical Society, Transactions*, 47:2, 1957. By permission of the author.

the Iberian peninsula. No carding was done until commercial cards were introduced in 1852.

Various spinning methods were used, depending on the area, but the drop spindle of the Mediterranean civilizations (and Peru, where it was used for wool) was unsuitable for short-staple cotton. Instead the point of a slender spindle was rested either in a small bowl or between the toes or on the ground. The whorls might have been shards of pottery, wooden disks, or simple crossbars. Warps were usually spun tightly, wefts somewhat more loosely and fully.

Precisely how the vertical loom originated in the Southwest can only be surmised. If the first heddle loom in the area was the backstrap loom, it was quickly discarded in the early historic era for the horizontal ground loom and vertical looms favored by the southwestern tribes. One can reason that the vertical loom evolved directly from the free-warp and Salish-type looms, or one can speculate—as some have done regarding the ancient Egyptians—that it was a solution to a problem of space. When the Pueblos brought cotton weaving into the *kiva*, their ceremonial chamber, perhaps they set the broad horizontal loom upright. Perhaps. The steps to the vertical Pueblo loom

—and hence the Navajo loom—ultimately remain a mystery.

Although three looms were used in the Southwest—horizontal, vertical, and backstrap—after 1100 the vertical loom dominated and became the one on which the Pueblos wove their finest textiles. Fig. 4-5 illustrates its construction. Note that the upper and lower bars, instead of being lashed to uprights, are attached to the ceiling and the floor.

The uppermost bar is permanently lashed to a ceiling beam. Just below, a thinner bar—the tension bar—is attached to the uppermost bar by a cord that spirals around both. As the name of the thin bar suggests, the cord is used to adjust the tension of the warp threads stretched below. This tension bar is in turn lashed with loops of cord to the upper bar (or warp beam) of the loom. The lower bar (or cloth beam) is held down by loom anchors buried in the floor. Occasionally, instead of several individual anchors a long, heavy beam was half-buried in the floor. Holes were bored at intervals along the protruding top edge, and through these holes cords cinched the lower bar firmly in place.

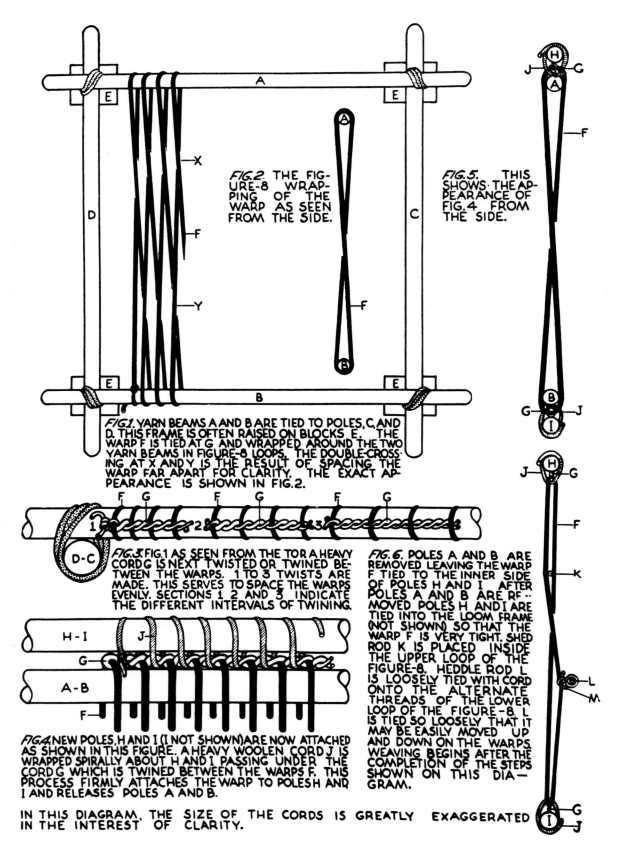

FIG.2. THE FIG-URE-8 WRAP-PING OF THE WARP AS SEEN FROM THE SIDE.

FIG.5. THIS SHOWS THE AP-PEARANCE OF FIG.4 FROM THE SIDE.

FIG.1. YARN BEAMS A AND B ARE TIED TO POLES C AND D. THIS FRAME IS OFTEN RAISED ON BLOCKS E. THE WARP F IS TIED AT G AND WRAPPED AROUND THE TWO YARN BEAMS IN FIGURE-8 LOOPS. THE DOUBLE-CROSS-ING AT X AND Y IS THE RESULT OF SPACING THE WARP FAR APART FOR CLARITY. THE EXACT AP-PEARANCE IS SHOWN IN FIG.2.

FIG.3. FIG.1 AS SEEN FROM THE TOP. A HEAVY CORD G IS NEXT TWISTED OR TWINED BE-TWEEN THE WARPS. 1 TO 3 TWISTS ARE MADE. THIS SERVES TO SPACE THE WARPS EVENLY. SECTIONS 1 2 AND 3 INDICATE THE DIFFERENT INTERVALS OF TWINING.

FIG.6. POLES A AND B ARE REMOVED LEAVING THE WARP F TIED TO THE INNER SIDE OF POLES H AND I. AFTER POLES A AND B ARE RE-MOVED POLES H AND I ARE TIED INTO THE LOOM FRAME (NOT SHOWN) SO THAT THE WARP F IS VERY TIGHT. SHED ROD K IS PLACED INSIDE THE UPPER LOOP OF THE FIGURE-8. HEDDLE ROD L IS LOOSELY TIED WITH CORD ONTO THE ALTERNATE THREADS OF THE LOWER LOOP OF THE FIGURE-8. L IS TIED SO LOOSELY THAT IT MAY BE EASILY MOVED UP AND DOWN ON THE WARPS. WEAVING BEGINS AFTER THE COMPLETION OF THE STEPS SHOWN ON THIS DIA-GRAM.

FIG.4. NEW POLES, H AND I (I NOT SHOWN) ARE NOW ATTACHED AS SHOWN IN THIS FIGURE. A HEAVY WOOLEN CORD J IS WRAPPED SPIRALLY ABOUT H AND I PASSING UNDER THE CORD G WHICH IS TWINED BETWEEN THE WARPS F. THIS PROCESS FIRMLY ATTACHES THE WARP TO POLES H AND I AND RELEASES POLES A AND B.

IN THIS DIAGRAM, THE SIZE OF THE CORDS IS GREATLY EXAGGERATED IN THE INTEREST OF CLARITY.

4-6: Details of warping the Pueblo loom. Denver Art Museum.

Warping was not done on this vertical arrangement but on two separate bars held parallel to the ground. The bars were usually supported at one end by two loom blocks (chunks of sandstone about a foot square with grooves cut to hold the bars) and at the other end by inserting the bars into holes cut for the purpose in the wall of the house. Four loom blocks or wooden beams with nails driven in a semicircle to secure the bar were sometimes used. Between the two bars the warp would be wound on in a figure eight.

Next, a two- or three-ply heading cord was twined between the warp threads along the outside edge of one of the bars (fig. 4-6), the number of twists between warps determining the spacing. The same was done at the opposite bar. An additional bar was then lashed through the heading cord at each end, and the initial warping bars were removed. As a result of twining the heading cords along the top and bottom of the warps, the completed fabric had four selvages—i.e., no warp fringe. The last step was to attach the warp to the rest of the loom (see fig. 4-5). The new lower bar was tied to the floor anchors, and the new upper bar was secured with a few loops of cord to the tension bar.

A few details remained before weaving could begin. Through one shed (formed by figure-eight warping) a shed rod was inserted; then the heddles and heddle rod(s)—the Pueblos used as many as five—were added. Finally, side selvage cords were tied to either the heading cords or bars. These extraheavy cords, twined around the weft as it made the turn to enter the next shed, provided the blanket with side selvages as sturdy as those at the ends. Just getting this far took the Hopi weaver a full day.

Additional tools included a batten that varied from 8 to 30 inches long and 1 to 3 inches wide and a comb (later supplanted by a kitchen fork). The weft was either wound into a small ball and inserted with the fingers or wound lengthwise along a slender stick. No specialized shuttle was used anywhere in the Americas. Some believe that this was fortunate, as a shuttle encourages monotonous weaving and unimaginative patterns. The shuttle increases speed but discourages elaboration of design.

The Pueblo weaver typically wove a band at the bottom of the loom and then turned the warp upside down and began at the opposite end (fig. 4-7), possibly to ensure that the two end borders matched. As the weaving progressed

4-7: Hopi blanket weaver, Oraibi, 1902. History Division, Los Angeles County Museum of Natural History.

out of reach, the weaver lowered the warp by loosening the tension bar and took up the slack by sewing folds of the completed cloth to the lower beam. The last few wefts had to be tediously darned in, first with a slender stick and finally with a needle from the yucca plant or later of steel.

With minor exceptions the Navajo loom was warped and operated identically to the Pueblo loom. And since the Navajos learned their craft from the Pueblos, it should be no surprise that their looms closely resembled the looms of their Pueblo teachers.

THE NAVAJO LOOM

The Navajos have always been better borrowers than innovators, but their textiles did not suffer from the lack of originality in their looms. Within a hundred years of their initiation into the art of weaving they surpassed their mentors as the major source of woven textiles in the Southwest. In 1795 Governor Fernando de Chacon wrote of the

Navajos, "They have increased their horse herds considerably, they sow much and on good fields; they work their wool with more delicacy and taste than the Spanish."

The Navajos derive from a mixture of ancestral lines that probably includes the Athapascan hunters from the North, the industrious Pueblo farmers, the warlike Utes, and desert wanderers skilled in basket making and scavenging. They arrived in the Southwest in small nomadic bands, picking up customs and physical traits as they encountered and mingled with other tribes. As a people their intrusion into the Southwest has been variously dated between A.D. 1100 and 1485, depending on which archaeologist or Navajo legend is consulted. Archaeological evidence favors the later date.

The historical beginning of Navajo weaving probably coincided with the Pueblo Rebellion of 1680. At that time the Pueblos joined forces to expel the Spanish intruder from their lands. The Navajos, relatively unmolested by the

4-8: Navajo loom. After Washington Matthews, *Navajo Weavers*, 1884.

4-9: Navajo loom. Santa Fe Railway Photo.

Spanish and aloof from the revolt, happily plundered both parties and enriched themselves with both Pueblo and Spanish sheep and horses. For twelve years the Pueblo people managed to keep the Spanish at bay, but in 1692 the intruders returned, crushing Pueblo resistance and forcing many to flee to the Navajos for refuge. Intermarriage took place to such an extent that today many Navajos exhibit physical characteristics closer to the Pueblos than to their Apache brothers.

Until the Pueblo Rebellion there is no record of Navajo weaving. The Navajos were never known to weave in cotton, and it was the Pueblo Rebellion that brought sheep in large quantities into Navajo possession. As a result of the Rebellion the Navajos remained under Pueblo influence for over seventy-five years—more than enough time for them to master the art for which they are now popularly known. Given that the Pueblo teachers were men, it is interesting that the Navajo women became the pupils. One authority

suggests that among the Athapascans women traditionally were the weavers, but it is also possible that Navajo men, traditionally hunters and warriors, simply didn't take to the sedentary art. Another authority states that the likelihood of Navajo women learning to weave from Pueblo men was so remote that the Navajo must have brought the tradition of the vertical loom with them from the Northwest Coast.

The first serious consideration of Navajo weaving was made by Washington Matthews in his report for the Smithsonian Institution in 1884. Matthews' illustration of the Navajo loom (fig. 4-8) shows its obvious similarity to the Pueblo prototype. The uppermost bar, however, instead of being attached to a ceiling beam, is lashed to uprights; and the lower bar, instead of being anchored to the floor, is held down with rocks or heavy logs. Even among the Navajos the loom frame varied considerably. The uprights might be two growing trees if they were the right distance apart; the lower beam might be lashed to the uprights; the uppermost

4-10: Navajo loom. Santa Fe Railway Photo.

4-11: Navajo weaver winding a warp, Chaco Canyon, New Mexico, 1904. Photograph courtesy of the Museum of the American Indian, Heye Foundation.

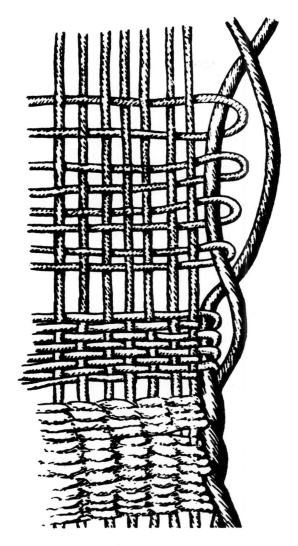

4-12: Diagram of Navajo selvage twining. From Charles Avery Amsden, *Navajo Weaving: Its Technic and Its History*, 1934.

edges. At the completion of the weaving the loose ends of the side selvage cords are either tied with the ends of the heading cords into corner tassels or darned back into the web.

The Navajos were also known to weave from both ends toward the middle, either by turning the entire warp upside down or by weaving down from the top. (Backstrap weavers from Peru and Mexico followed the same practice—see Chapter 5.) In addition to achieving symmetrical borders this technique produced a stronger top edge. As the end of the weaving approached, whether it was in the middle or at the top, the decreasing shed space left no room for the heavy beating in that the blanket received elsewhere. (According to Matthews, "It is by the vigorous use of the batten that the Navajo serapes are rendered waterproof.") The last few inches were thus apt to be somewhat looser in texture, and placing this area away from the ends probably increased the overall life of the textile.

The minor differences between Pueblo and Navajo weaving techniques are disputable—for example, whether or not the Navajos inserted the weft with their fingers or used a stick shuttle (and if so, of what length) or whether it was characteristic of the Navajos to weave a small portion at the top of the cloth before completing the weaving from the bottom up. The chances are that the Navajo experimented with many of the Pueblo techniques, adopting some completely, using others occasionally, and rejecting a few completely.

One technique that they rejected was weaving their blankets from selvage to selvage. While the Pueblos shifted position to weave their broad *mantas* straight across, the Navajos sat in one spot and worked as much as they could reach before moving to the adjacent area. The blankets thus woven exhibited what came to be called "lazy lines" where the areas came together (fig. 4-13).

bar might rest in crotches (fig. 4-9); the loom might be located either indoors or out, depending on the weather (fig. 4-10).

The warp is wound over two bars in a figure eight identical to the Pueblo method (fig. 4-11). But the bars, instead of being supported by holes in the wall of a room or by four blocks, are tied to side poles—the distance between them conforming to the desired length of the blanket—and the whole rectangular frame is supported off the ground. The rest of the loom preparation follows the Pueblo method: the warps are spaced apart by twining heading cords between them outside the bars (fig. 4-6); the shed rod and heddles are inserted with the aid of the cross maintained by sticks inserted into the figure eight; side selvage cords are fastened to the lower bar, drawn up, and tied loosely over the upper bar. The cords are twisted with the passage of the weft through them (fig. 4-12), thus strengthening the

The course that Pueblo and Navajo weaving took in America was largely determined by white colonization. Contact with technologically advanced society inevitably eroded native art and crafts. By the early 1880s railroads had penetrated the Southwest, bringing with them commercial cloth, commercial dyes, and Germantown yarns. The Pueblo weavers, who generally did not weave for sale but for their own use, found these new materials cheaper and easier to use than those of their own manufacture, and weaving, already on the decline, began to die out.

The Hopi Pueblos, farthest removed from both the Spanish traders and the path of the railroad, escaped the assimilating influences somewhat better than most. On the whole they were not particularly interested in trading with the whites, and today they are the only Pueblo people to continue the cultivation of cotton—but even they use it mainly for ceremonial clothing. Southwestern loom weav-

4-13: Navajo blanket, Hudson Bay style, 1880–1890, showing lazy lines. Los Angeles County Museum of Art, Georgia O'Keeffe Collection.

4-14: Navajo blanket, Serape style, 1850–1865. Part of the weft is of unraveled baize, dyed red with cochineal. Anthropology Division, Natural History Museum of Los Angeles County.

ing formerly embraced the following pueblos, or tribes: Zuni, Hopi, Acoma, Santa Clara, Nambe, Cochiti, Navajo, Pima, Papago, Maricopa, Tepehuares, Opata, Tarahumare, Yaqui, Mayo, Cora, and Huichol. Today, a mere three hundred years since the white man set foot in the Southwest, weaving is limited to the Navajo, the Hopi, and to some extent the Zuni Indians.

The course of Navajo weaving has been divided into periods that differ in name and span according to the commentator. The Classic Period was reached around 1850–75, the time of their greatest technical proficiency. During this period the famous *bayeta* blankets were woven from unraveled English baize—or *bayeta* in Spanish—a flannel cloth brought in through Mexico (fig. 4-14). Although baize was manufactured in many colors, the Navajos preferred red to the virtual exclusion of all others. Handspun wool, often spun three times to acquire the desired strength and thinness, formed the warps of these blankets.

Between 1863–68 Kit Carson rounded up the Navajos for the U. S. government, and they were interned at Bosque Redondo, a miserable "round grove" of cottonwood trees outside Fort Sumner in east-central New Mexico. By 1875 weaving was on the decline. A Transition Period from 1875–90 is noted, during which Navajo weaving might have died out altogether were it not for the tourist trade that followed the course of the railroads west. Germantown yarns and analine dyes replaced bayeta and hand-dyed indigo yarns, and cotton twine threatened to supplant handspun wool warps. The ready-made yarns gave the Navajos time to experiment with their weaving, and during this period they produced what one expert has called "eye dazzlers," blankets swimming in color (fig. 4-15). They stopped weaving blankets for their own use, instead buying trader's blankets from the East, and devoted all their weaving time to items for sale. The tourist interest gradually caused them to switch from weaving blankets to rugs.

The Rug Period, 1890–1920, was the nadir of Navajo weaving. Standards dropped precipitously. In 1910 the government, hoping to improve the quality of Navajo meat, introduced the Rambouillet sheep. If the mutton was better, the wool was worse. Unlike the Spanish *churro* sheep, a common sheep from Spain with long, straight, and almost greaseless wool, the Rambouillet sported a short, crimpy, oily fleece. Whereas the *churro* wool could be woven without washing, the wool of the tastier sheep was at once difficult to wash, card, and spin. By 1920 weaving had reached its all-time low.

A revival followed, with renewed interest in native vegetable dyes and traditional designs, and today Navajo weaving survives as a kind of endangered species. The Navajo historically has shown great talent in learning from and adapting to alien cultures while retaining tribal identity. The transition from blanket to rug might have concluded an art form, but it also created a new basis for economic and social stability. The forces for assimilation have been blunted in the past decade by a reawakening of ethnic pride in America, but the question remains of how long the economics of Navajo life will support tribal pride and tradition without vitiating the best in Navajo art.

4-15: Navajo "Eye-Dazzler" blanket, 1880–1890. The range of aniline dyes includes brown, light brown, gray, orange, red-orange, and yellow. Los Angeles County Museum of Art and Anthony Berlant, Santa Monica, Ca.

5. The Backstrap and Other Primitive Looms

The loom is after all only the frame upon which a principle, weaving, is worked out, and . . . there is considerable reason for the supposition that it may have been invented more than once. —H. Ling Roth

Who first twisted the delicate fiber into strong, continuous thread? The race has yet to build a monument to him whose genius first guided the pliant thread into warp and weft, and to the silent myriad millions who added to his original conception. —M.D.C. Crawford

If present knowledge is any indication, the monument, if and when it is built, may well commemorate *her* (not his) genius. Men may have helped with its construction, but, judging from most aboriginal societies, it seems to have been women who first manipulated the threads on the primitive loom.

The word "primitive" carries an unfortunate suggestion of crudeness that might be applied by association to the textiles woven on them. This would be a mistake. One might just as easily label as "crude" the sculptor's chisel. It is almost an irrelevant observation, for the artistry of the sculptor resides in the hand and in the eye, not in the chisel, and to a large extent the same could be said of the weaver. The more primitive the tool, the more it demands of its user. The very crudeness of her loom lends credibility to J. Alden Mason's comment in *The Ancient Civilizations of Peru* that "in the textile industry the Peruvian woman is considered by many technical experts to have been the foremost weaver of all time." More so even than the Chinese drawloom weaver, with his elaborate contrivances for lifting individual silk threads in the warp.

Yet the designation "primitive" must be made, for we are considering the tool and not its product. The backstrap loom is by all definitions a primitive tool: when the finished cloth was removed from it, nothing remained but a handful of sticks, a belt of some kind that went around the weaver's back, and a few lengths of string. Contrast this to the modern loom, with its machined ratchet wheels and pawls, its steel heddles, harnesses, and reeds, and its bulky frame that does not depend on the fabric to give it shape.

It is incredible how little mechanical help early weavers actually needed to produce magnificent textiles. Not all cultures wove cloth, nor did all those that did weave produce magnificent textiles, but it should not be forgotten that some of the world's finest weaving was made on a tool as crude as the backstrap loom.

THE BACKSTRAP LOOM

ORIGINS

The term *backstrap loom* (also variously known as the girdle-back, hip, waist, belt, or stick loom) refers to any loom on which the warp is stretched between some stationary object and the body of the weaver (fig. 5-1). The weaver provides the tension by leaning back against some kind of belt—of wood, leather, fabric, or cordage—that is attached to both ends of the cloth beam. At the opposite end the warp beam might be tied to a stake or tree, a hook in the wall or ceiling of a house; or it might be supported between two uprights or even held firm by the soles or toes of the weaver's feet (fig. 5-2). The warp beam in fact might be omitted altogether and the warp ends simply knotted in a bunch about a hook or post (see fig. 5-26).

Dr. Junius Bird calls the backstrap loom—in spite of its apparent simplicity—a complex device and cautions against the assumption that people arrived at it easily as a solution to weaving. In a sense it is a finer and more responsive instrument than the modern treadle loom, because the warp tension is constantly tuned by the weaver. The backstrap weaver is herself a part of the loom, a part that the modern loom does not possess: an "automatic warp regulator."

Although perhaps most closely associated with the Indians of Peru and Mexico, this loom has first cousins all over the world, particularly in remote areas of Asia such as northern Hokkaido in Japan, in Malaysia, Indonesia, and the Philippines, in China, Korea, Tibet, Burma, and parts of India. It is still the common loom in southern Mexico, the highland regions of Guatemala, and part of the Andes and until recently was popular among the Navajo and Zuni Indians of Arizona for weaving belts, sashes, and other narrow fabrics.

5-1: Basic backstrap loom. A weaver of blue cotton *rebozos*. Acatlán, Guerrero, Mexico, 1964. Photograph by Donald Cordry.

5-2: Montagnard weaver, Vietnam, 1968. Photograph by Howard Sochurek.

Where it came from and when it arrived at these various places may never be determined satisfactorily, for in most areas of the world the climate has not favored the preservation of textiles. And as for preserving the loom, even if the wooden bars managed to survive the ravages of time and the weather, the backstrap loom lacked most distinguishing features. With the exception of the sword beater common to many kinds of primitive looms at best one might find notches near the ends of the beams to keep the belt or cord from slipping off or perhaps carvings or other decorations affixed to the ends—all in all, not much to go on.

In eastern Asia the earliest archaeological evidence of a loom dates from a Bronze–Iron Age culture on the southwestern frontier of the Western Han empire (206 B.C.–A.D. 8). At a site in Shizhaishan, Yunnan Province four bronze loom parts (cloth and warp beams, shed stick, and sword beater) have been excavated that probably belonged to a ceremonial backstrap loom. The lid of a cowrie container from the same site shows some miniature bronze figures engaged in a ritual offering, six of whom are weaving on foot-braced backstrap looms similar to that shown in fig. 5-2. Where this type of loom originated and whether it was the oldest type of loom in China remain unknown, but recent examples of the foot-braced backstrap loom have been found in several of the Southeast Asian countries and offshore islands. Particularly well documented is the Atayal loom of Taiwan (see fig. 5-25).

Tracing the diffusion of a loom type presents enormous difficulties, well typified by Saul H. Riesenberg's analysis of the shape of loom bars to determine how the backstrap loom spread through the Caroline Islands of the Pacific: "The loom with cylindrical beams diffused to the end of the Caroline chain. Later, open-warp weaving was introduced on Yap, but did not pass beyond this northwesterly Caroline Island. The board-shaped beam spread independently and later into the Carolines, replacing the round beam in most islands, existing side by side with it in some of the eastern islands, changing to the form square in cross-section in some of the central islands. Meanwhile, the older, cylindrical beam spread south, probably via Kapingamarangi, into Melanesia. The square beam of the central islands also diffused south, again via Kapingamarangi, so that both round and square beams are found in Melanesia today. The board-shaped beam failed to reach Melanesia except for the single occurrence at St. Matthias, which must represent a separate introduction." And this does not take into account the diffusion of single- or double-edged sword beaters, single or double lease sticks, and other differences found in the Carolines. Some design elements were probably of native origin, too.

Riesenberg believed that loom weaving probably spread from some Southeast Asian source down through Indonesia into the Carolines and then into northern Melanesia. Two kinds of backstrap loom have been noted as common in Indonesia: one that uses a continuous, cylindrical warp that is shifted around the end beams as weaving progresses (the loom found in the Carolines) and one that uses a discontinuous warp that is wound on a board-shaped warp beam and unwound as more warp is needed (see fig. 5-33). The latter loom also employs a reed beater in front of the heddles. At least one authority doubts 'that the second derived from the first and even questions whether there might not have been a third, even simpler loom in use at an earlier stage. The diversity of opinion among ethnologists makes interesting reading. One thinks that the earliest Indonesians didn't weave but wore leaves or tapa cloth; another believes that the backstrap loom originated in Indonesia and migrated across the Pacific to South America and elsewhere. The evidence behind these theories is scanty and open to broad interpretation.

Although nature has rewarded the textile archaeologist in few areas of the world, she has been truly munificent in one—the dry coastal plain of Peru. From ancient cemeteries in the desert textiles thousands of years old have been unearthed looking just as fresh and as colorful as the day they were woven. It was the custom to bury along with the worker the tools and accoutrements of his or her profession. Consequently, the graves have yielded not just the clothes in which the weaver was buried but workbaskets filled with spindles and bobbins (fig. 5-3), balls of yarn, and lengths of fabric—and sometimes actual looms.

The earliest Peruvian woven textiles discovered thus far are some pieces of plain, weft-faced weaving found at Guitarrero Cave and dated 5780 B.C., long before the hunter-gatherers settled down to till the soil. (One expert dates earlier material c. 8600 B.C.) Whether or not these fragments were woven on a backstrap loom cannot be determined, but the backstrap loom has generally been acknowledged as the earliest *known* Peruvian loom. The evidence supporting this belief rests partly on the fact that most of the textiles found have measured less than thirty inches in width—about as far as a backstrap weaver could reach from selvage to selvage—and partly on early representations of looms, the earliest of which appears on a Mochica pottery vessel (fig. 5-4) dated by various experts from 200 B.C. to A.D. 1000.

Along with the plain weaving found in Guitarrero Cave were many examples of twined bags and basketry that James M. Adovasio and Thomas F. Lynch find similar to twined examples from North American sites—such as Danger Cave, Utah; Fishbone Cave, Nevada; and Fort Rock Cave, Oregon—that date as far back as the ninth millenium B.C.! Adovasio and Lynch conclude from these discoveries that "in South America, as in North America, twining is the oldest and most basic textile making technique from which most, if not all the others, are ultimately derived." This opinion enlarges what had been the prevailing view, as expressed by Dr. Junius Bird, that weaving derived from "the experimental manipulation of yarn and that until someone invented the heddle, weaving was of very minor importance."

The looms pictured on the Mochica pot do not possess heddles, but opinions differ as to what this might signify. Some believe that the uneven fell of the cloth indicates that the weavers were doing tapestry, not selvage-to-selvage weaving, and thus did not need heddles. Others attribute the missing heddles to the artistic license of the potter. It is also possible that at the time when this pot was made heddles were not yet in general use.

The heddle loom is thought to have been introduced to Peru along with pottery by a culture outside Peru sometime between 1500–1200 B.C. Although weaving was done earlier, it seems to have been only one of several fabric techniques of minor importance compared to twining. For example, from the Chicama Valley in northern Peru some seventy-eight percent of discovered fabrics, dated c. 2500–1200 B.C., were twined, while less than four percent were woven. The balance consisted of knotted netting and looping. Some pieces that contained both twining and weaving techniques (see fig. 1-4) suggest that weaving (probably without heddles) might have been used to patch

5-3: Peruvian workbasket and spindles from grave. Courtesy of the
American Museum of Natural History.

5-4: Backstrap weavers shown on the lip of a Mochica pottery bowl,
Chicama Valley, Peru. Reproduced by courtesy of the Trustees of the
British Museum.

holes in twined cloth. After about 1200 B.C. the ratio of twining to weaving is reversed. The suddenness of this change, along with the appearance of skilled pottery, indicates that both pottery and heddle-loom techniques may have arrived with outside intruders.

Early heddle-loom weaving was probably no more elaborate than twining. The changes came slowly, but in the Early Horizon Period (900–200 B.C.) textile technique and design took off. By the end of this period Peruvian looms were turning out tapestry, double cloth, gauze weaves, and pattern weaves as well as fabric for painted cloth and embroidery—mostly of textiles less than 30 inches wide woven on the backstrap loom.

Mostly but not all. Some Peruvian fabrics have been found with widths up to seven and eight feet, and one fabric mentioned in a Peabody Museum report of 1938–39 measured 47 feet long and 12 feet wide. For many years this puzzled archaeologists. The obvious limitation of the backstrap loom is the width of the fabric that can be woven on it, about 23 to 30 inches, the distance that a weaver could reach. The standard explanations until recently have been that either narrow pieces had been stitched together or a Navajo-type frame loom was used, but in 1947 Truman Bailey discovered a third possibility—the three-person backstrap loom (fig. 5-5). The three-person loom might suggest an example of primitive technology gone berserk, but the loom was probably no more cumbersome than the two-person warp-weighted loom, particularly for tapestry in which the weft was not carried from selvage to selvage (though the widest fabrics found so far are plain weave).

THE PERUVIAN LOOM
The diagram in fig. 5-6 illustrates the simplicity of the basic Peruvian backstrap loom. The heddle bar (c) has been raised to open the shed, and the weft, wound on the bobbin

5-5: Three-person backstrap loom, Peru. The loom was made by Truman Bailey from directions given by one of the weavers, who had seen or heard of such a loom from her mother. Courtesy of Truman Bailey and the American Museum of Natural History.

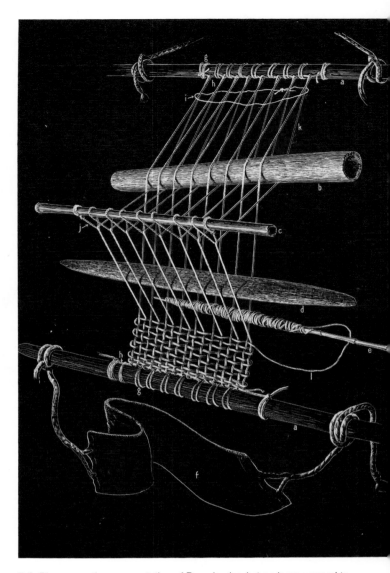

5-6: Diagrammatic representation of Peruvian backstrap loom warped to produce a fabric with four finished selvages. (a) Loom bars. (b) Shed rod. (c) Heddle rod. (d) Batten. (e) Bobbin. (f) Backstrap. (g) Warp lashing. (h) Heading string. (i) Lease or laze cord. (j) Heddle leash cord. (k) Warp. (l) Weft. Courtesy of the American Museum of Natural History.

(e), has been partially inserted. After beating down this shot of weft the batten (d) would be removed, the heddle bar lowered, and the opposite shed opened by the shed rod (b). The weaver would insert the batten into this narrow opening and, turning the batten on edge, widen it enough for the passage of the weft.

The yarn was wound in a figure eight around two stakes and then transferred to heading cords, as on the Navajo loom (see Chapter 4), which in turn were lashed to the loom bars. Heading cords gave the option of weaving cloth with four finished selvages; the weaver had only to turn the loom around after starting one end and to begin again at the other, finishing where she left off, near the original cloth beam. Although heading cords were the usual arrangement, in some instances the warp was wound directly around the loom bars themselves.

Peruvian cotton, probably a hybrid of a wild variety, had been cultivated since 3600 B.C. and perhaps even earlier. It provided the basic textile fiber until about 1000 B.C. when the alpaca and llama were domesticated. The guanaco and vicuna, although never domesticated, also furnished wool, and that from the vicuna was particularly prized for its silkiness. It has been said that the Peruvians spun bat hair as well, but most modern authorities discount this as an exaggeration. Bat hair is too short to spin—though human hair is not and occasionally was used.

The Peruvian weaver spun yarn on a long, thin spindle that often doubled as a bobbin after the spinning was done. It was usually pointed at both ends with a terracotta, bone, or metal wheel in the center, both to add momentum to the spin and to anchor the spun yarn. The Peruvian spinner, like the spinner of ancient India, rested the lower end of the spindle in a clay bowl that may have contained water to moisten the fibers or fingers. (Some claim that damp fiber spins better than dry, but the Zoque Indians of Central America dipped their fingers in ashes while spinning—to dry them.) The yarn produced by this method varied from heavy to fine, and the finest was two to three times finer than modern machine thread made from the same material—up to 210,000 yards to the pound.

Textile technology in Peru evolved further during the Early Intermediate Period (c. 200 B.C.–A.D. 600), with new techniques added during the Middle Horizon Period (A.D. 600–1000). By the time of the Incas (A.D. 1100–1532) all their textile techniques had been fully explored and developed. The Incas, heirs of a developed technology, changed only the textile designs, and even these were based on those that they inherited.

Some Peruvian techniques were so extraordinary, either in their ingenuity or in their perfection of execution, that it would not be just to conclude a discussion of the Peruvian backstrap loom without mentioning a few of the most outstanding examples. Their tapestries, for example, some with 260 to 280 picks per inch, far excelled the finest Gobelin works from Europe, which rarely exceeded 80 wefts per inch. The Gobelin masterpieces contained less than half the number of warp threads per inch as their contemporary Peruvian counterparts. The backs of Peruvian tapestries were so carefully finished that they were difficult to distinguish from the fronts. Backs of the Gobelins, by contrast, were a mass of tangled yarns.

It might be unfair to carry the comparsion further, because the Peruvian and Gobelin tapestries were woven for different purposes. But detailed comparisions are not necessary. Consider the Peruvian technique of *weft interlocking*, also called weft scaffolding (fig. 5-7), an ingenious device to change the warp color in mid-textile. Tem-

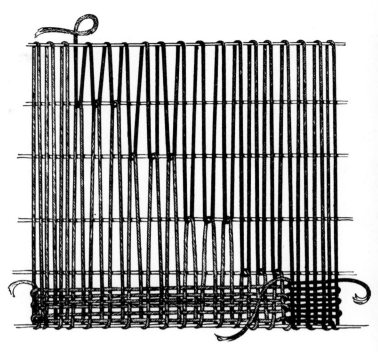

5-7: Diagram of weft scaffolding from the Paracas Necropolis, Peru. Detail of mantle in multicolored patchwork technique, a variant of plain weave, approximately 6″ square. Published in 1942 by The Regents of the University of California, reprinted by permission of the University of California Press.

porary scaffolding wefts to which new warps were either attached or looped were tied across the loom. After the weaving was completed, the scaffolds could be removed from interlocking warps or not, as the weaver desired. If the new warps were dovetailed around the scaffolding weft, the scaffold could not be removed without the fabric separating at the juncture.

Another Peruvian innovation was the textile shaped on the loom (fig. 5-8), a garment that was woven to fit the wearer without tailoring. Various techniques were devised. For example, by adding double warps through a looped end of a single warp the weaver could broaden—or by reversing the procedure taper—the textile. Other techniques for shaping included angling one or both loom bars, adding extra wefts, or adjusting warp tension.

Among the stunning catalog of fabric techniques (including plain weave, tapestry, double and even triple cloth, warp and weft patterning, warp and weft interlocking, gauze, weft pile, brocading, embroidery, appliqué, featherwork, twining, wrapping, looping, sprang, braiding, macramé, ikat, and tie-dyeing, among others) are found Peruvian twills—relatively few during the early periods but more later on. Lila M. O'Neale, an authority on ancient Peruvian textiles, suggests three ways in which twills might have been woven on the backstrap loom.

The first and most time-consuming method was to pick up one warp at a time, either with the fingers (or a warp picker) or by darning in each time a thin sword for opening

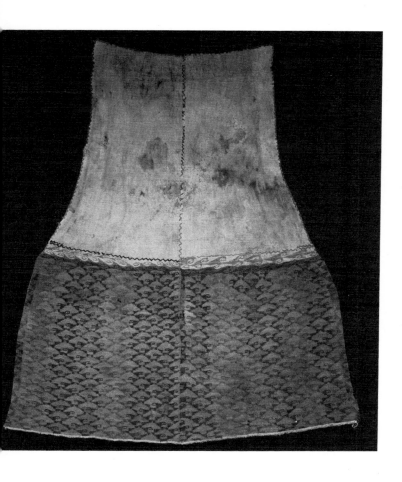

5-8: Peruvian textile. Shaped weave with tapestry section. Courtesy of the American Museum of Natural History.

5-9: Backstrap loom with pattern sticks, from near Lima, Peru, showing partly finished web of double cloth. The rods above the heddle rod separate the warps for the pattern. Courtesy of the American Museum of Natural History.

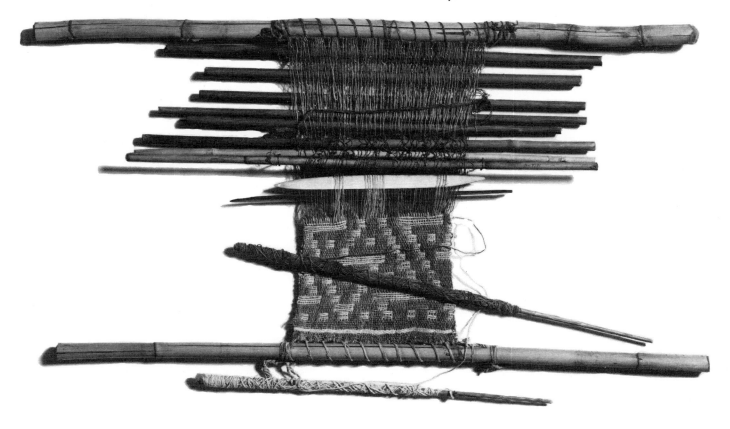

the shed. The second method, a slight improvement over the first, was the use of *pattern sticks* (fig. 5-9) to separate the warps for each shed. With this technique small sticks or swords still had to be darned into the warp, but the entire pattern could be set in place before weaving began. After each pick the weaver would move the lowest pattern stick up to the top in the same position that it occupied at the bottom, and the pattern would automatically be repeated after each full cycle of the rotating sticks. This was an extremely laborious system for wide fabrics having many warps per inch. (Using pattern sticks on the backstrap loom was a popular technique throughout Southeast Asia, the Philippines, Melanesia, Polynesia, and Indonesia as well, where up to sixty sticks might be employed for one pattern [fig. 5-10].) A third method was the use of *multiple heddles*—at least two plus a shed bar for a 2/1 twill.

THE MESOAMERICAN LOOM

A similar story of the backstrap loom is repeated in Mesoamerica (the area spanning Central Mexico south to Nicaragua), but that story begins about 1,500 years later. The earliest evidence of plain weave in Mesoamerica appears on pottery shards from Tehuacan dated c. 1800–1200 B.C. Actual textile fragments dated only slightly later than this indicate the use of cotton for one set of weaving elements (and yucca fiber for the other) at about this time. Some authorities now believe that descendants of the same groups from the flood plains of the Amazon basin or northern South America that brought ceramics and weaving techniques to the eastern slopes of the Andes migrated north as well, reaching the southeastern edge of the Mexican plateau around 1500 B.C.

On the basis of admittedly scanty evidence some authorities suggest that Mesoamerican textiles were influenced by southern cultures twice: between about 2000 and 1500 B.C. when loom weaving and cotton fibers were introduced and again after about A.D. 900 when a wider range of techniques was utilized for the first time. Mesoamerican weavers, although possibly infused with Andean techniques, had developed their own traditions based on the narrower range of fibers and dye shades available to them. Their textiles, for example, emphasized twills, warp floats, gauzes, and other warp-oriented techniques in contrast to the weft floats and exuberant tapestries of the South. Even though the Andean weavers experimented more widely with technique, 'design, and color (O' Neale distinguished 190 dye hues from textiles at the Paracas Necropolis), Mesoamerican weaving was equally competent and vital within the more limited confines of its own tradition.

According to Indian traditions, weaving in Mesoamerica originated with the gods. In Yucatan the Mayans attribute it to the wife of a god, worshipped under the name of Ix-azalvoh. The Toltecs credit their god Quetzalcóatl (also

5-10: Backstrap loom with 54 pattern sticks, Badja people, Indonesia. Courtesy of the American Museum of Natural History.

5-11: Backstrap weaver from Mendoza Codex. Mother instructing her 14-year-old daughter at the loom. From Kingsborough, *Antiquities of Mexico*, 1831–48.

responsible for picture writing and the smelting of gold), and the Aztecs look to Xochiquetzal, goddess of flowers and craftsmen. More scholarly sources theorize that the Aztecs learned weaving from earlier residents of Mesoamerica, who picked it up from the Mayans, and so on back to the earliest sources of diffusion from South America.

Whatever the source, by 1520 when Cortéz marched into the Aztec capital of Tenochtitlan (now Mexico City), cutting short both the career of Montezuma II and the history of the Aztec Empire, the backstrap loom (fig. 5-11) must have been in widespread and active use. The tribute roll of the *Codex Mendoza*, which dates to the time of Montezuma II (1502–1520), states that every eighty days the Aztecs collected from their conquered foes over 1,328,000 cloaks, 72,000 *maxthal* (skirts), 96,000 *huipiles* (tunics of skirts), and 4,000 bales of cotton and much *henequen* (hemp) as raw material for the Aztec weavers.

Many of the traditional shapes and designs of Mesoamerican textiles have survived into the modern era. One Preconquest garment, the *quechquemitl*, was usually formed by joining one end of each of two woven rectangular strips to the side of the other (fig. 5-12). The result was a neckpiece that was popular among the Otomi, the Aztecs, the Totonacs, and the Tepahuas. The Otomi, however, developed a method of weaving the strips so that they formed not rectangles but curved shapes that fell gracefully over the shoulder (fig. 5-13). The curves were woven in on a backstrap loom by using part of the warp as weft (fig. 5-14), a technique whose ingenuity certainly

rivals ancient Peruvian methods of making loom-shaped textiles.

Although difficult to understand without a step-by-step—really a thread-by-thread—explanation (see Bodil Christensen's article in the Bibliography for more detail), figs. 5-15 and 5-16 give some idea of what takes place. The loom in fig. 5-16 is set up to weave both halves of the *quechquemitl* at once (unlike the loom in fig. 5-14), and one-half—at the top of the loom—is completed. The white warp is cotton, and the black is red wool. At the bottom the red warps have been cut and are hanging loose in readiness for use as wefts. Not counting the preparation of the yarn or the setting up of the loom, the weaving of an Otomi *quechquemitl* takes a good weaver two days. The embroidered joining of the two halves requires a third day.

Other garments took longer. A Zoque Indian *huipil* (a knee-length tunic), for example, might take two months of steady work if the design were complicated. Donald and Dorothy Cordry wrote of a Zoque weaver in Ocozocoautla: "Because almost all the weaving was *'con labor'* she wove from sunrise to sunset, and all her meals were brought to her at her loom. These designs are so complicated and difficult that one can think of nothing else while weaving, and for this reason most women do not make them after they are married." Very little of the ancient *huipil* weaving is found today; machine-made clothes are cheaper—though less durable.

The contemporary Mexican backstrap loom—whether Otomi, Zoque, Huichol, or Zapotec—has changed little since ancient times. The same can be said of Guatemala

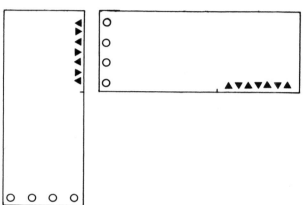

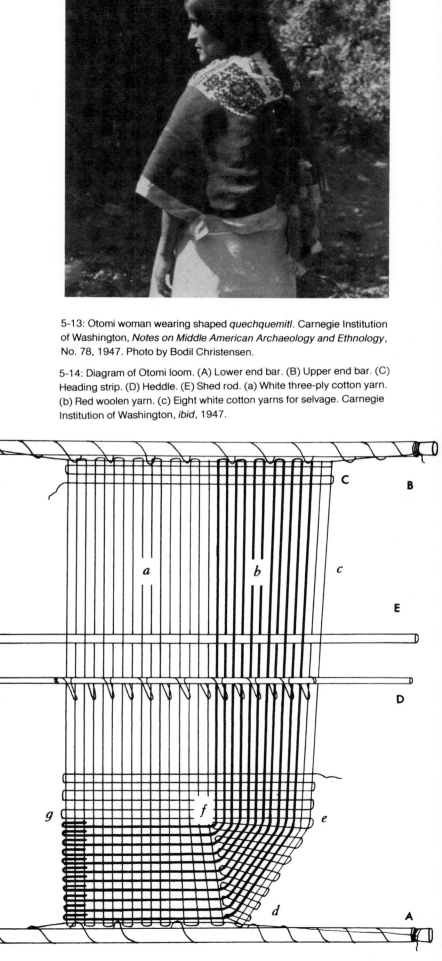

5-13: Otomi woman wearing shaped *quechquemitl*. Carnegie Institution of Washington, *Notes on Middle American Archaeology and Ethnology*, No. 78, 1947. Photo by Bodil Christensen.

5-14: Diagram of Otomi loom. (A) Lower end bar. (B) Upper end bar. (C) Heading strip. (D) Heddle. (E) Shed rod. (a) White three-ply cotton yarn. (b) Red woolen yarn. (c) Eight white cotton yarns for selvage. Carnegie Institution of Washington, *ibid*, 1947.

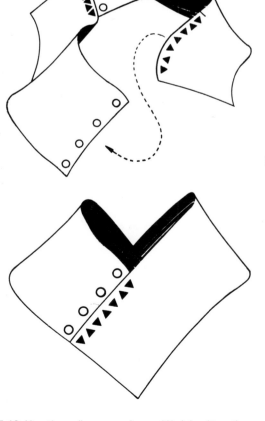

5-12: How the ordinary *quechquemitl* is joined together. Drawings by Andy M. A. Chowanetz.

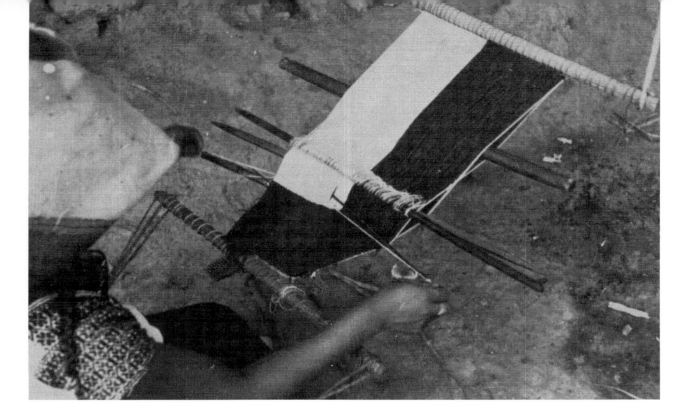

5-16: Woman weaving at Otomi loom. Carnegie Institution of Washington, 1947. Photo by Bodil Christensen.

5-15: Otomi loom with fabric half woven. Carnegie Institution of Washington, 1947. Photo by Bodil Christensen.

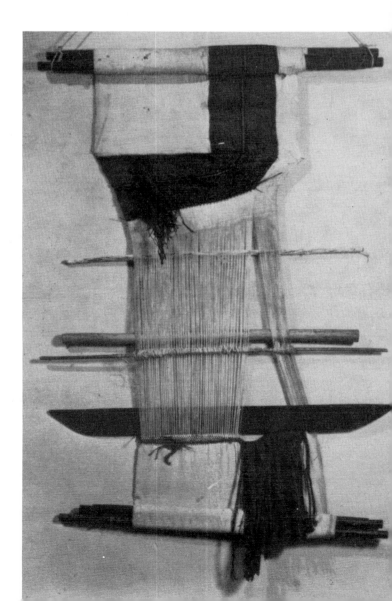

where weavers work *en palitos* ("on little sticks"), the number of *palitos* depending on the complexity of the pattern. The Zoque loom (fig. 5-17) as used in Tuxtla Gutiérrez, Mexico, is in·most ways typical of all. The smaller sticks—the heddle bars, shed bar, and bobbin—are of smooth bamboo; the end bars might be made of any heavier wood close at hand; the batten, the one part of the loom about which the weaver feels possessive—perhaps because more than any other part it controls the quality of the work—is usually fashioned of Brazil wood. In Guatemala some weavers put pebbles in the hollow bamboo shed rods to hear the pleasant rattle as they worked. It is possible that the rattling served some superstitious or spiritual purpose as well. (Clappers, rattles, and gongs—devices to make noise with each passing of the weft—have also been noted in Indonesia and Sri Lanka.)

The Mexican loom is warped much like the Navajo loom, with the warp fastened not around the end bars but around

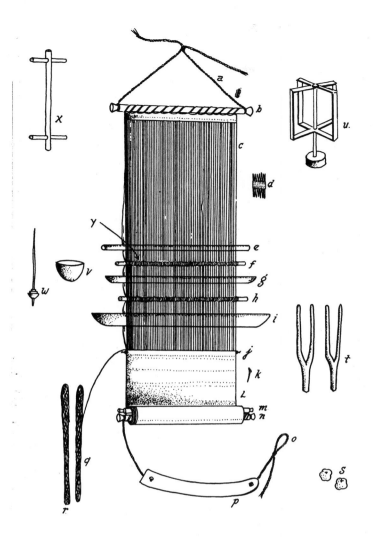

heading cords or loom strings (fig. 5-18). After the warp is wound, it is placed around two "rolling sticks," or temporary end bars, and separated into sections. A cord is then tied to one end of the stick that will become the true end bar. The cord is inserted through the warp, tied to the other end of the bar, and spiraled back around the end bar lashing the cord and groups of warp threads to it. Lastly, the "rolling stick" is removed, and the same procedure is followed at the opposite end.

The heddles are applied by the same method that is followed for the Navajo loom. The weaver faces the loom in the weaving position and lays on the right side of the loom a ball of string that contains more than enough material for the heddles. She passes the end of the string to the left through the shed, leaving the ball in its original position, and ties a loop at the end large enough to admit the heddle rod. She passes a straight, slender rod (the heddle rod) horizontally through the loop until the point of

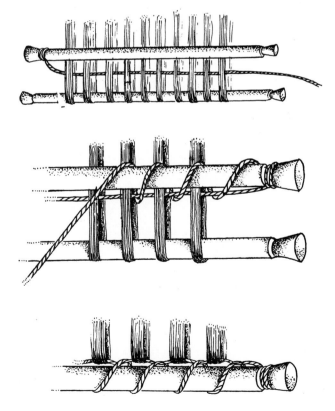

5-18: Details of the warping process on the Zoque loom. Courtesy of Southwest Museum, Los Angeles.

5-19: Navajo woman weaving a belt. After Washington Matthews, *Navajo Weavers*, 1884.

the stick is even with the third thread of the warp. She pulls out a fold of the heddle string through the space between the first and third threads, twists it around, forming a loop, and pushes the point of the rod to the right through this loop. She continues forming loops and advancing the rod from left to right until each of the anterior (alternate) warp threads of the lower shed is included in a loop. When the last loop is made, she ties the string firmly to the rod near the right end.

The Zoque weaver wove a small heading strip at one end (see fig. 5-17), turned the loom over, and began again at the opposite end. The heading strip, also seen in Pueblo and Navajo weaving, regulated the spacing of the warp threads and stabilized the width of the cloth. It also kept the last few picks of the weft, where the cloth might be weakest, away from the end selvage. (The Zoques had a saying that if one part of the cloth was looser than another, it was a sign that the intended recipient of the cloth was one "of evil life.")

THE AMERICAN SOUTHWEST LOOM

The backstrap loom made its way north from Mesoamerica into the Pueblo region of the American Southwest, probably around A.D. 700–1000. It competes with the horizontal ground loom for the honor of being the first loom to penetrate into that region, but both gradually yielded in popularity to the two-bar vertical loom. The principle of weaving was the same on both backstrap and vertical looms: they were warped, the heddles added, and the cloth woven in much the same way. But two variations on the basic Peruvian and Mesoamerican backstrap loom turned up in Pueblo America.

One was a warping variation that permitted the weaving of tubular, usually warp-faced, belts (fig. 5-19). A similar method was used by certain South and Mesoamerican tribes, but the method differed from the tubular weaving of the Scandinavians, the Salish Indians, and various other South American tribes in which the warp ends were looped around a common transverse cord to form a removable joint (if the cord were pulled out, the warp loops separated and the "tube" could be opened up flat [see fig. 3-16]). The Pueblo tubular warp was wound in continuous, spiral loops (without the figure-eight cross) around the end bars so that the entire warp could be rotated as weaving progressed. To keep the threads in order (fig. 5-20), the top set of warps was divided into alternate sections, and a cross kept with two sticks tied together at the ends (d). The three rods (c) were laced through both the upper and lower warps to keep them from rotating under the pressure of beating in the wefts. The upper of the two remaining rods (e) is the shed rod, the lower the heddle bar.

The second variation, the rigid, or reed, heddle, was believed to have been introduced to the New World by Europeans, though its origin is unknown. Early examples have been found in the Roman culture, but the earliest illustration of a rigid heddle occurs in fourteenth-century Europe. The Zuni and Hopi Indians have used this type of loom (fig. 5-21), probably acquired from the Spanish, but apparently found it less efficient than their own heddle rod and string-loop heddles. And no wonder—the rigid heddle was heavier and took up more room than the traditional heddle rod, and it probably took longer to thread as well.

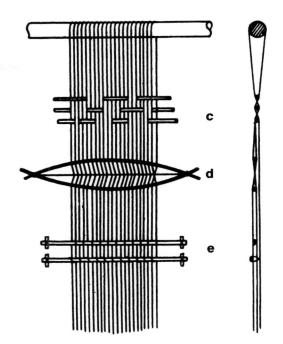

5-20: Diagram of Zuni loom for belt weaving. Reproduced by permission of the American Anthropological Association from the *American Anthropologist* 26 (1): 74, 1924.

5-21: Zuni woman weaving a belt on the rigid-heddle loom. After Washington Matthews, *Navajo Weavers*, 1884.

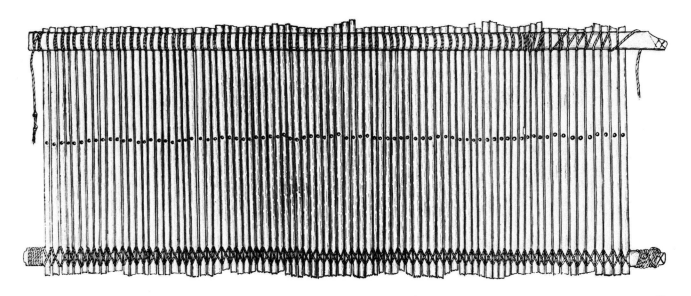

5-22: Pueblo rigid-heddle frames. After Otis T. Mason, *A Primitive Frame for Weaving Narrow Fabrics*, 1901.

5-23: Pueblo batten knives. After Otis T. Mason, *A Primitive Frame for Weaving Narrow Fabrics*, 1901.

The rigid heddle substitutes for both the shed bar and the heddle rod. It consisted of a series of short reeds, or sticks, held parallel by top and bottom cross bars (fig. 5-22). Each reed had a small hole bored through its center, and the warp yarns passed alternately through these holes and the slots between each reed. When the heddle was pushed down, one shed was formed; raising the heddle created the countershed. One set of warps, those passing between the reeds, remained stationary, while the yarns passing through the eyes of the reeds slid up and down beside them, changing the sheds. The picks were usually inserted without the aid of a bobbin and were beaten in either with the edge of the hand or with a special sword beater (fig. 5-23).

Where and when the rigid heddle originated remain a mystery. It was used in ancient times in Scandinavia and

5-24: Finnish rigid-heddle loom. Heddle is carved from a single piece of wood. Photograph by István Rácz.

possibly as early as the Swiss Neolithic Age. The earliest representation of a rigid heddle comes from an early fourteenth-century manuscript. The typical northern European version was carved from a single piece of wood (fig. 5-24), but for centuries French and Spanish peasants made it from reeds. The Pueblo rigid heddles were crude devices by comparison, twelve to thirty inches long, fashioned from mesquite sticks lashed together with leather thongs. Although variations on the rigid-heddle loom have been found from Bolivia to Canada and from the American Southwest to Indonesia, China, and Japan, it seems to have been a poor adaptation for Pueblo weavers, who abandoned it altogether early in the twentieth century.

THE ATAYAL LOOM
The foot-braced backstrap loom (see fig. 5-2) appears to be a particularly primitive weaving instrument because the length of the cloth that it can weave is limited by the length of the weaver's legs. Although the loom is uncommon, ethnologists have reported its use by certain tribes in South America and a number of tribes in Southeast Asia and the Pacific from as far west as the Assam highlands to as far east as the St. Mathias Islands in the Bismarck Archipelago. The loom lacks any reed or temple to regulate the width of the fabric. The utter simplicity of the tool requires the utmost skill of the weaver.

All foot-braced looms use a spirally wound cylindrical warp. The shed making and weaving take place only in the top layer, and, as the fabric enlarges, the entire warp is shifted around the end bars to bring more warp within reach. When the beginning meets the end, the weaver cuts the warp apart for a garment fringed at both ends. Its length is, of necessity, about two yards—twice the length of the weaver's legs.

On the island of Taiwan Atayal weavers have devised a unique modification to this loom that permits a longer cloth to be woven (fig. 5-25). The extended loop of warp passes around the breast beam, back under the top two layers of the warp, and under the main beam to an extra warp beam pegged to the ground somewhere beyond. This unusual

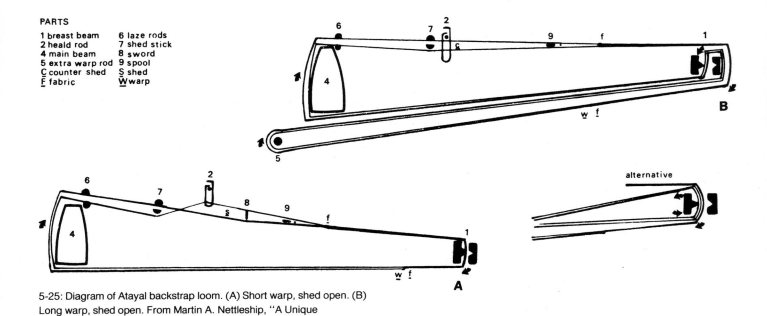

PARTS

1 breast beam   6 laze rods
2 heald rod   7 shed stick
4 main beam   8 sword
5 extra warp rod   9 spool
C counter shed   S shed
F fabric   W warp

alternative

5-25: Diagram of Atayal backstrap loom. (A) Short warp, shed open. (B) Long warp, shed open. From Martin A. Nettleship, "A Unique South-East Asian Loom," *Man* (V) 1970, 686–698, fig. 3.

adaptation kept the same working surface before the weaver yet more than doubled the capacity of the loom.

The diagram also illustrates that the breast beam consisted of two parts—a tongue-in-groove combination that clamped the warp (or cloth) securely so that the cloth wouldn't slip during the beating in. To clamp the two parts together, the weaver wrapped the cords of the backstrap around extensions at the ends of both parts of the breast beam. To advance the warp, she simply loosened one end, pulled the fabric through the slot, and rewound it shut.

While not typical of all foot-braced looms, the main warp beam of the Atayal loom was large and drumlike. Its size gave the feet more purchase against the beam, and its hollowness produced a reverberating drumbeat with each thud of the beater against the fell of the cloth. Perhaps, like the rattling shed rods of certain weavers in Guatemala, this served some ritualistic purpose. It may have been the desire to retain the same drumbeat that led Atayal weavers to

such an unconventional method of lengthening the warp rather than simply pegging the main warp beam to the ground or some other support, as did the Iban or Dusan weavers in Borneo (see fig. 5-30).

THE AINU LOOM

On the northern island of Hokkaido, Japan an aboriginal tribe still weaves on backstrap looms that probably resemble the earliest looms of Japanese origin. The Ainu loom (fig. 5-26) includes the customary cloth bar, to which a backstrap of wood or bark is fastened, but instead of a warp bar the warp ends are bunched together and tied to a stake in the ground. Thus, if the Ainus used only a shed rod and heddle, the warp threads would not lie parallel but would form an isosceles triangle.

Ainu weavers applied two solutions to the problem of nonparallel warps, both versions of the *osa*, or warp spacer. The reed warp spacer (fig. 5-27), probably in-

5-26: Ainu backstrap loom. From David MacRitchie, *The Ainos*, 1892.

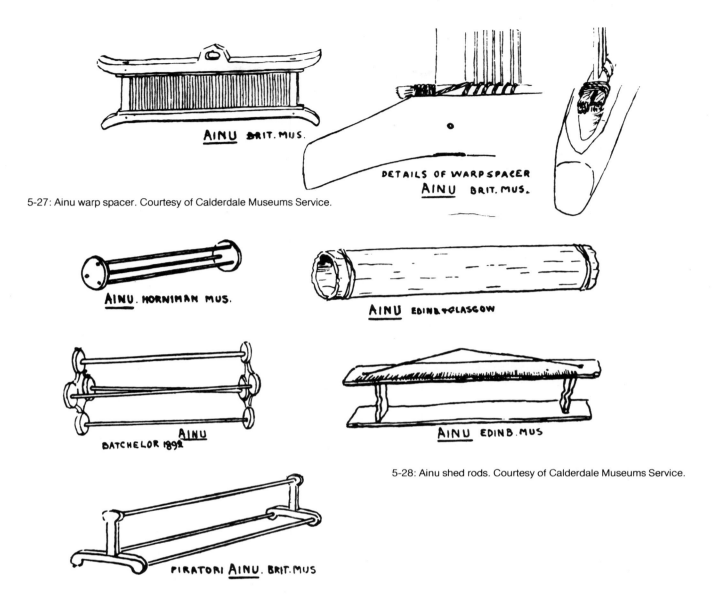

5-27: Ainu warp spacer. Courtesy of Calderdale Museums Service.

DETAILS OF WARP SPACER AINU BRIT. MUS.

AINU. HORNIMAN MUS.

AINU EDINB + GLASGOW

AINU BATCHELOR 1892

AINU EDINB. MUS

5-28: Ainu shed rods. Courtesy of Calderdale Museums Service.

PIRATORI AINU. BRIT. MUS

troduced from the outside, resembled the modern reed except that it was not used for beating in the weft. Broad wooden knives did the beating in, knives that were similar to the Pueblo beaters but broader. The warp spacer was placed *behind* the heddles to align the warp threads parallel for weaving. With the warps properly spaced and aligned, weaving could proceed with a shed rod and heddle arrangement. The Ainu shed rod (fig. 5-28) was often constructed of two or more smaller rods or pieces of wood. The second solution, a combination shed rod and warp spacer (fig. 5-29), served the double function of spreading the warp threads and maintaining one of the sheds. H. Ling Roth believed that this version, appearing more primitive in construction, preceded the reed spacer and was probably indigenous to the Ainu. Such a device, he reported, has been found on no other primitive loom.

The fiber woven on the Ainu loom was drawn from the inner bark of the elm tree. (The Ainus were far from unique

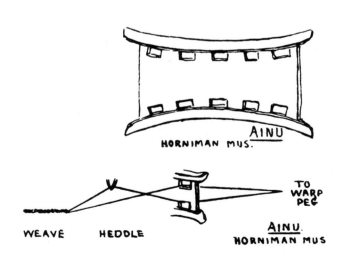

AINU HORNIMAN MUS.

WEAVE   HEDDLE   TO WARP PEG   AINU. HORNIMAN MUS

5-29: Ainu combination shed rod and warp spacer. Courtesy of Calderdale Museums Service.

in their unusual choice of fibers: the Santa Cruz islanders, also backstrap weavers, wove garments of banana fiber, and, the reader will recall, the Chilkat Indians twined their blankets with shredded cedar bark. Other North American Indians used inner bark for fabrics as well.) The elm strands, softened and separated either by chewing or by immersion in water, were not spun but were tied end to end. Great care was taken while weaving to keep all the knots on the underside of the cloth; such painstaking diligence is said to have limited progress on a fifteen-inch-wide fabric to about a foot a day.

The Ainus were not the only people to utilize a reed spacer. The sign for the reed dates back to the Middle Kingdom in Egypt. By Coptic times weavers used a reed constructed of flat iron wires set in a wooden frame. The reed probably developed first as a warp spacer and made the transition to batten quite late in the history of the loom. Roth illustrates with several examples from Indonesia and the Philippines how this transition might have occurred. (It must be said, however, that while the modern reed-batten probably did evolve from the early reed warp spacer, Roth's microcosmic genealogy must be regarded as an illustrative, not a definitive, example. Too few details of the reed's ancestry are known to construct an accurate family tree in a broader context.)

1. The Iban or Dusun loom (fig. 5-30) uses no reed at all. A warp beam is lashed across two uprights, and the shed on the continuous warp is changed with a shed rod and single heddle rod.

2. The Igorot loom from the Philippines (fig. 5-31), also using a continuous warp, is similar but with a reed. The fine reed ends are set in a groove in the cane base and secured with fiber twined between the dents and around the reeds. At the top the reeds fit loosely into a slot in another cane bar but are not secured. The reed as constructed is not strong enough for beating in and was probably used only for spacing the warp. A sword beater, further evidence of the reed's sole function as warp spacer, is also shown with the loom.

3. The Ilanun loom fron Sabah (formerly North Borneo) used a reed (fig. 5-32) somewhat sturdier in construction, though still rather loose at the top.

4. To complete the sequence, Roth cities next a Javanese reed that is rigid at top and bottom. The Javanese or Balinese loom (fig. 5-33) clearly shows the reed (H) placed in front of the heddles where it can be used as a beater.

The reed-batten as we know it today, either suspended from above or pivoting from below, required the development of a loom *frame*, a relative latecomer to weaving history usually associated with the development of foot treadles and counterbalance harnesses. (This broad generalization admits of many exceptions, such as the African toe-treadle loom, which supports both the reed and harnesses from a simple, movable tripod [see fig. 6-22].) The notched supports for the revolving Javanese warp beam suggest the embryonic stages of a developing loom frame. But here we begin to encroach upon the subject matter of Chapter 6, "The Treadle Loom," where we will

5-30: Iban woman weaving on two-bar backstrap loom. From Hose and McDougall, *Pagan Tribes of Borneo*, Vol. I, 1912, by permission of Macmillan, London and Basingstoke.

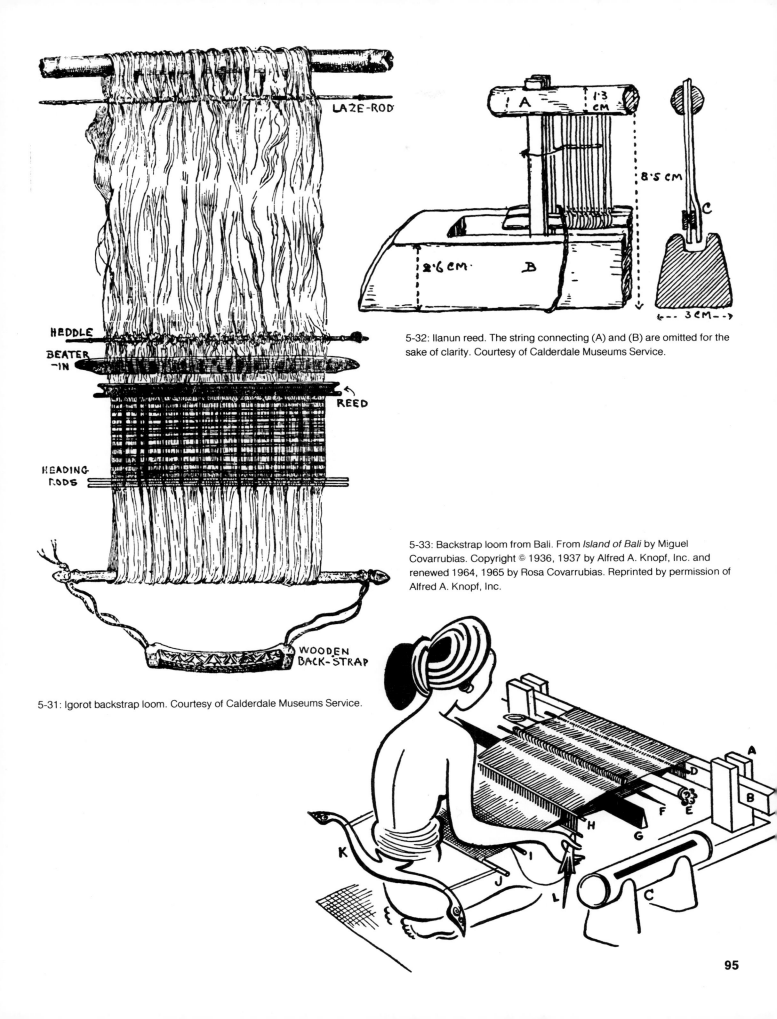

LAZE-ROD

HEDDLE

BEATER-IN

REED

HEADING RODS

WOODEN BACK-STRAP

5-31: Igorot backstrap loom. Courtesy of Calderdale Museums Service.

A

1·3 CM

8·5 CM

2·6 CM.

B

C

3 CM

5-32: Ilanun reed. The string connecting (A) and (B) are omitted for the sake of clarity. Courtesy of Calderdale Museums Service.

5-33: Backstrap loom from Bali. From *Island of Bali* by Miguel Covarrubias. Copyright © 1936, 1937 by Alfred A. Knopf, Inc. and renewed 1964, 1965 by Rosa Covarrubias. Reprinted by permission of Alfred A. Knopf, Inc.

A

B

C

D

E

F

G

H

I

J

K

L

find the backstrap loom in combination with treadles, heddle harnesses, and a frame that must be distinguished from those presented here as "primitive."

OTHER PRIMITIVE LOOMS

The category "other" is reminiscent of that response solicited on questionnaires, usually prefaced by (e), into which one's opinion or explanation invariably falls. The simple statements (a) through (d) just do not suffice, and the same must be said of primitive looms. There are simply too many different kinds and not enough known about the relationships, if any, among them. This section, then, represents a necessarily incomplete sampling, exhibiting looms of special mechanical or historical interest independent of their position in the overall history of the loom.

THE BENT-STICK LOOM

Two looms from the Solomon Islands illustrate the two-bar loom in its simplest form. The Nissan Island loom (fig. 5-34) consists of a single stick about 43 inches long, split lengthwise and propped open by two smaller sticks, each

about four inches long. The ends of the loom are tied to check further splitting. A warp of bast fiber is wound in a continuous fashion around the center of the loom. The longer of the two pointed sticks inserted in the weaving could be a shed rod. There is no heddle: the opposite shed is opened by darning in the shorter stick after every second pick. The weft is inserted by means of the fingers and a needle. The Bougainville Island loom (fig. 5-35), manipulated in a similar manner, stretches the warp by the spring tension of a bent stick. In Roth's opinion these two looms probably descended from the sophisticated Solomon Islands art of mat weaving.

Another bent-stick loom is used in Guiana, on the northern coast of South America, for weaving bead aprons (fig. 5-36). In the Río Ucayali region of northern Peru and northwestern Brazil a similar loom is formed by bending a piece of cane or liana into an oval and securing the ends. Two parallel rods or cords or one of each are fastened across the oval and function as the end bars. Other looms from the same area are made in triangular, pear, and wishbone shapes. The textiles woven on these looms and on analogous looms from other areas such as Colombia

5-34: Fine mat loom without heddle, Nissan Island. Approximate length from beam to beam is 4". Courtesy of Calderdale Museums Service.

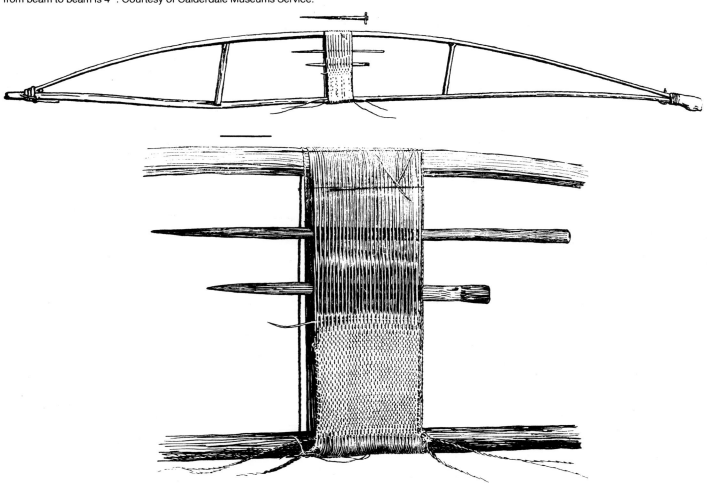

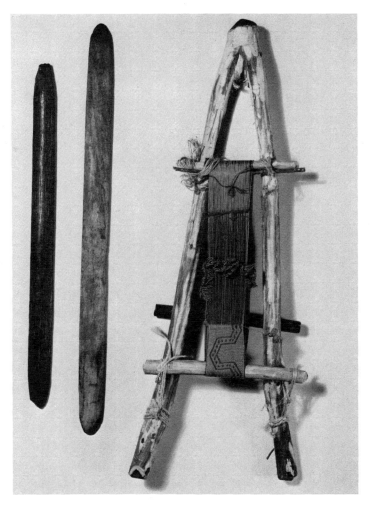

5-35: Bent-stick loom, Bougainville Island, width approximately 14″, used for making armbands. Inv. No. Vb 8317; North Solomon Islands, Bougainville, Buin; collection Felix Speiser, 1930. Museum für Völkerkunde und Schweiz.-Museum für Völkerkunde, Basel.

5-36: Loom with partly woven bead apron, British Guiana. The Museum of the American Indian, Heye Foundation.

5-37: Forked frame loom for weaving belts. Nazaret, Alta Quajira, Colombia, 38″ high. Courtesy of the American Museum of Natural History.

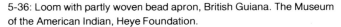

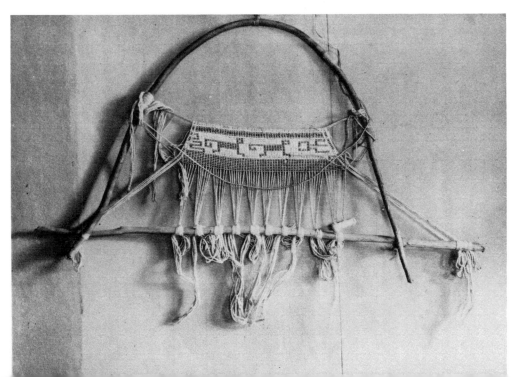

5-38: Navajo belt loom, 1910–14. The Museum of the American Indian, Heye Foundation.

(fig. 5-37) and the American Southwest (fig. 5-38) are necessarily limited to narrow bands for sashes, belts, saddle girths, headbands, and so on.

Perhaps the most curious example of the bent-stick belt loom is one that Roth attributes to "the North American Slave Indians" (fig. 5-39). The bowed stick stretches a row of sinews (Roth calls these the "*pseudowarps*") a distance of about two feet from their fastening at the folded leather patch to the opposite tip of the bow, where they are tied. The pseudowarps are held parallel by a pseudowarp spacer, a piece of birch bark perforated with a line of holes through which the sinews are threaded. The sinews, despite the appearance of a true warp, merely provide a suspension system for the true warp, which is wound around them. With the pseudowarps in place, a pair of red-stained sinews is twined across them just in front of the folded leather. The true warp, drawn from a continuous spool of clear sinew, is then wound at right angles around the pseudowarps, and the wefts are introduced. They are porcupine quills of various colors that are inserted from underneath (between the rows of pseudowarps) and bent over and under the true warps. The warps are then pressed back toward the leather apron, thus aligning and hiding the lower warps under the upper warps. Altogether a very ingenious piece of work!

THE ARAWAK LOOM
In South America Lila M. O'Neale has identified three general types of primitive looms: the Peruvian, or backstrap, loom that introduced this chapter; the Río Ucayali, or bent-stick, loom with its wishbone or other related-shape variants; and the Arawak, or Amazon, loom. The Arawak loom is used by widely separated people in the central and northern parts of the continent, mainly for narrow fabrics. It resembles the backstrap loom without the backstrap. Instead of being held between a post and the weaver's back the end bars are supported by two upright poles. By winding the warp continuously around the bars the loom can produce a seamless tube; if the warp is wound around a removable transverse cord (fig. 3-16), it can produce a four-selvage cloth twice the length of the loom.

The usual fiber was cotton, both white and brown, grown on cotton bushes that in pre-Columbian times reached up to fifteen feet in height. Most Indians ginned the cotton with their fingers, though some squeezed out the seeds between wooden rollers. After ginning the cotton was cleaned and fluffed up for spinning. Some Indians—the Mundurucú, Piro, and Central American women—accomplished this by beating it with sticks (fig. 5-40); others—the Carajá, Guató, Churapa, Guarayú, Chacobo, Guaraní, and Guana—used the bow introduced by West European missionaries, snapping the bowstring in the cotton to separate the fibers (fig. 5-41).

At least a dozen different tribes in South America wove

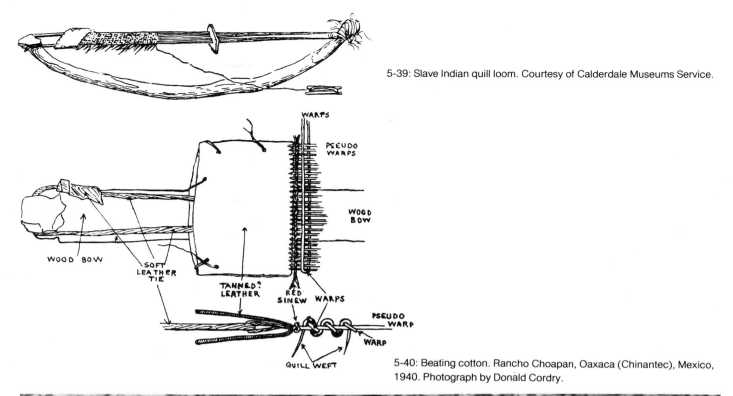

5-39: Slave Indian quill loom. Courtesy of Calderdale Museums Service.

WARPS

PSEUDO WARPS

WOOD BOW

WOOD BOW

SOFT LEATHER TIE

TANNED? LEATHER

RED SINEW

WARPS

PSEUDO WARP

WARP

QUILL WEFT

5-40: Beating cotton. Rancho Choapan, Oaxaca (Chinantec), Mexico, 1940. Photograph by Donald Cordry.

5-41: Cleaning cotton with bow. From Charles Knight, *The Pictorial Gallery of Arts*, 1845.

their cotton or wool into a wide, seamless tube called a *tipoy*. The wearer stepped into it and either folded the top half down over a belt at the waist or left it up as a pouch in which to carry a baby. South American weavings were not generally tailored into clothes but were worn more or less as they came off the loom.

THE ARAUCANIAN LOOM

In central Chile the Araucanians wove wool—first of guanaco but later, after the Spanish conquest, of sheep —into belts, shirts, ponchos, and breech-cloths. They probably learned the art from northern neighbors during pre-Inca times. Their loom (fig. 5-42) is reminiscent of the Pueblo loom, a simple rectangular frame lashed together, only simpler. The yarn is warped directly onto the loom frame, which is laid flat on stakes for that purpose about

twenty inches above the ground. Two women, sitting at opposite ends of the frame, warp the loom by rolling a ball of yarn back and forth to each other along the ground. The loom is then lifted upright and tipped so that the uprights (*huicha huichahue*—"with both feet on the ground") lean against the wall or roof of the house. The heddle bar rests on two thinner longitudinal bars that are tied to the end bars. Two smaller bars, lashed across the uprights, form the lease sticks, or *rañinelhue* ("Lord of the Center"). As the weaving progresses, the bottom beam is loosened, the cloth rolled up upon it, and the beam retied further up the uprights. At the halfway point the weaver turns the loom over and finishes the fabric from the other end.

As with many primitive cultures, the availability of cheap factory goods has been responsible for the deterioration of Araucanian weaving. In some tribes contact with western civilization had an even more shattering effect, destroying the unity and homogeneity of primitive cultures. Those crafts that were embedded in religious ritual, as weaving often was, inevitably suffered as the culture and ritual splintered and collapsed.

Contact with western civilization often breaks the continuity of primitive culture. This break interrupts the step-by-step process of invention. It is this slow, deliberate process that leads many to conclude that the loom must have been invented over and over again by many primitive societies, each responding to a need by inventing and then improving upon a way to weave some kind of fiber into some kind of cloth. Despite contact with an outside culture primitive societies often resist the introduction of new ideas. Thus the circle closes; this resistance to outside influences reinforces the belief that the loom had many local origins.

The deterioration of primitive weaving is inevitable as the world grows smaller and factory-made goods become increasingly available to remote societies. It would seem to be only a question of time before observations such as the following, made by R. Goris in 1956, would reflect a world that the traveler would no longer be able to visit: ''Every Balinese girl must learn to weave and in every house there is at least one loom. Particular proficiency or artistry is admired. With about 200,000 looms Bali could even be called one big weaving mill.''

The primitive loom served well the society that did not consider speed a factor in textile production. In general it offered limited mechanization and maximum flexibility. It provided the least possible interference between the manipulated fiber and the hand of the artist. As the loom developed, mechanization relieved the weaver from much of the work previously done with the fingers or, in the case of the backstrap loom, the body. With so much of the weaver already involved in the process of weaving, it was inevitable that sooner or later her feet would be called upon to do more than their most active task thus far: holding the warp beam on a backstrap loom. The role of the feet and the looms employing them are the subject of the next chapter.

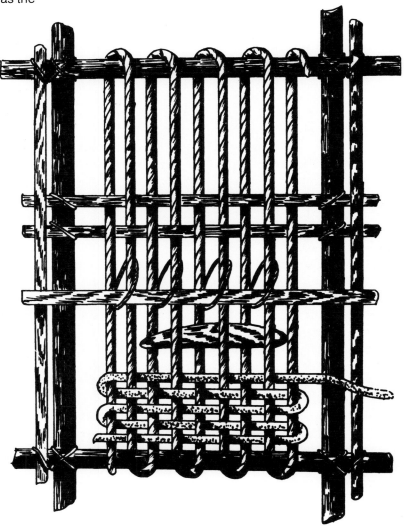

5-42: Diagram of Araucanian loom. Courtesy of Ciba-Geigy Ltd.

6. The Treadle Loom

Practically all cultures represent a sort of equilibrium between man's needs and inventiveness on the one hand and the demands and opportunities of the environment on the other. —Roderick U. Sayce

The philology of such words as shawl, carpet, chintz, calico, gauze, bandanna, and satin, and the plan of the eighteenth century loom, plainly show the Asiatic origin of our textile industries. —M.D.C. Crawford

The feet didn't have to wait for the invention or dissemination of treadles to find something to do at the loom. Certain Asian weavers used their feet to support the warp bar on the backstrap loom (see fig. 5-2). In the Palestine area Arab weavers on the horizontal ground loom employed a foot much as a hand might be used to control the sword beater (fig. 6-1). It is not as if the toes were untalented appendages. The skill with which an Arab woman could spin with her toes (fig. 6-2) should leave us all wondering what we may have lost by encasing our feet in leather.

What brought the feet into more active participation in weaving was undoubtedly the necessity for more "hands" at the loom. The addition of foot treadles for the purpose of changing sheds freed the hands for the more delicate work of inserting the weft and beating it in. The hands, of course, had always performed these tasks but were interrupted in their handling of the weft by having to attend to the alternation of sheds in the warp.

This apparently did not make much difference until silk and fine cotton were developed as fibers for more delicate weaving. At that point Asian weavers must have discovered that these fibers required more careful handling than was possible on their present loom, probably a variation of the backstrap loom. What the precise chain of events was is impossible to say, but most experts believe that the desire to weave fine fabrics of silk and cotton created the need for—and thus the invention of—the treadle loom.

The *treadle loom*—the phrase suggests a single loom or a product of a single, traceable ancestry. Would that it were so clear! The phrase is used here to indicate any type of loom that employs the foot (or feet) to change sheds by a treadling action (fig. 6-3), regardless of whether the foot actually depresses a pedal or pulls on a cord that is looped over the ankle or toe. This includes a variety of looms from different, though parallel, families.

The developed horizontal treadle loom appeared in Europe about the year A.D. 1000. The earliest surviving illustrations (see fig. 8-4) show an already perfected invention with revolving cloth and warp beams, treadles that were connected to counterbalanced heddle harnesses, and possibly even a suspended reed-beater. The frame, though rendered as a rather flimsy structure, is certainly recognizable as a forerunner of the modern horizontal floor loom. While little is known about the origin of this medieval European loom, most experts believe that it was descended only indirectly from the early East Asian progenitors.

A sequence of development is not easily established. It is not enough to say that looms evolved as increasingly sophisticated tools and then to list them in order from simple to complex. All handlooms, no matter how primitive or sophisticated, involve four processes that are subject to varying degrees of complexity or mechanization: (1) a system for holding the warp threads parallel, (2) a means of forming alternate sheds, (3) a process for inserting the weft, and (4) a manner of pressing it home. The processes developed in different cultures at various rates. In some places, for example, an integrated loom frame developed in advance of treadles and heddle harnesses; in other places it was the other way around. The evidence that we have consists of mere film clips from a long documentary. We can't splice them together end to end, but we can look at the segments that exist and perhaps speculate as to what has been lost to history.

THE LOOM FRAME

The frame of the contemporary treadle loom is a compact and efficient structure (see fig. 8-31). It supports the warp and cloth beams; the reed beater; the heddle harnesses, treadles, and various lams, pulleys, or jacks that connect them; the ratchets and pawls; the levers and brakes that control the advancing of the warp; and occasionally a

6-1: Weaver holding sword beater with toes as she beats in weft on Palestinian ground loom. Photograph by Shelagh Weir. Reproduced by courtesy of the Trustees of the British Museum.

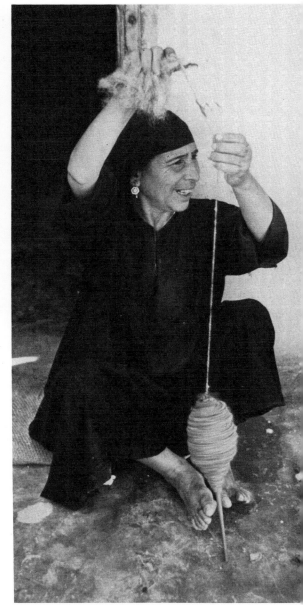

6-2: Arab woman spinning with toes. Photograph by Shelagh Weir. Reproduced by courtesy of the Trustees of the British Museum.

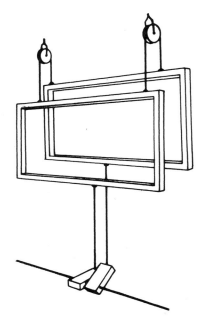
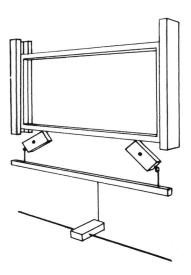
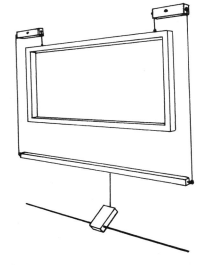

6-3: Diagram of treadling action. Drawing by Andy M. A. Chowanetz.

6-4: Bail backstrap loom with slotted warp posts. From Jasper, *De Inlandsche Kunstnijverheid in Nederlandsch Indie*, 1912.

footrest or even a bench for the weaver. In essence the treadle loom has not changed since the Middle Ages. Precisely how all those elements came together in one structure may never be known, but some clues are provided in H. Ling Roth's conjectures on the development of the loom frame in Malaysia and Cambodia.

Roth sees the rudimentary beginnings of a loom frame in the warp posts of the Malaysian or Bali backstrap loom (fig. 6-4). The slotted posts, fastened to heavy wooden feet, supported a flat warp beam around which the extra warp length was wound. Because of the shed rod and single heddle-rod arrangement no further supporting frame was required.

The flat warp beam, which seems characteristic of that part of the world, is found on several other looms with partly developed frames. On a Cambodian loom (fig. 6-5)

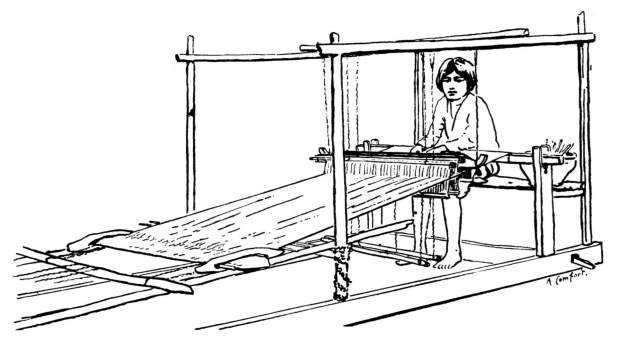

6-5: Cambodian loom. Courtesy of Calderdale Museums Service.

104

6-6: Malaysian loom with flat warp beam. Courtesy of Calderdale Museums Service.

the beam is supported by slotted end pieces that are in turn held in place by cords tied to still another bar. This latter bar is itself secured by ropes to a stake or some other stationary object, out of the picture to the left. The significance of this drawing, in Roth's opinion, lies in demonstrating that the complete loom frame may have developed from two distinct parts—the portion that supports the reed and heddle harnesses (perhaps borrowed from China) and the part that supports the warp beam.

The Malaysian loom (fig. 6-6) illustrates a further possible development, still using the flat warp beam. Here the warp-beam supports have joined the main body of the loom frame. Note that the heddle action is governed by whippletrees, whereas the system on the Cambodian loom appears considerably more rudimentary. There the cords supporting one heddle frame seem to be carried over a cross bar at the top of the loom and down to the second heddle frame. The heddles operate on the counterbalance principle but without the use of pulleys or whippletrees. Yet the drawing is not definitive.

Roth's analysis implies that the frame developed a step at a time as was needed to support such increasingly sophisticated moving parts of the loom as heddle frames, revolving warp beams, and suspended reed beater. The truly significant advance, of course, was not from the flat warp-beam support to the full frame but from the rod heddle to the suspended heddle frame and treadle arrangement. Some authorities think that the missing link between the rod-heddle loom and the two-treadle loom as discussed above was probably a *one*-treadle loom. Such a

loom did exist in China and Japan (see fig. 6-14), but whether it occupied that intermediate stage of development and was a necessary antecedent to the two-treadle loom is not known.

What casts some doubt on the role of the one-treadle loom as an essential evolutionary step to the two-treadle loom is the utter simplicity of some of the earliest known two-treadle looms, such as the Indian pit loom. It seems pure conjecture to suppose that the idea of raising one set of heddles through a fixed shed by treadling *had* to precede the idea of raising and lowering two heddle frames by an apparatus as simple as that discussed below.

THE PIT TREADLE LOOM

THE INDIAN LOOM
The pit treadle loom is the earliest known treadle loom to be used in India. It is conceivable that a single-treadle loom preceded it, but no evidence to that effect has yet been discovered. According to some experts, the Indian loom probably changed but little in four thousand years.

The loom, as described in Mill's *History of British India*, consisted of two bamboo rollers, one for the warp and one for the cloth, and a pair of harnesses. The shuttle, somewhat longer than the width of the warp, also served as the batten. The weaver suspended the harnesses from a branch of a tree under which he had dug a pit to accommodate his legs and the "treadles" of the loom. These treadles consisted of a loop tied at the bottom of a string attached to each harness into which the weaver inserted

105

6-7: Indian pit loom. From Baines, *History of the Cotton Manufacture in Great Britain*, 1835.

6-8: Indian pit loom with roller beams. From Gilroy, *The History of Silk, Cotton, Linen, Wool and Other Fibrous Substances*, 1845.

his big toes for changing the sheds. The breast beam was pegged to the ground and the warp was stretched out its full length, which, as Mill stated, "makes the house of the weaver insufficient to contain him. He is therefore obliged to work continually in the open air; and every return of inclement weather interrupts him (fig. 6-7)."

The weaver could advance the warp and roll up the finished cloth by loosening the rope that was fastened to a stake behind him. Other historians dispute Mill's description and placement of the pit loom. Alfred Barlow, for example, who must have fancied himself a most keen observer, insists in *The History and Principles of Weaving by Hand and by Power* that the loom was always set up under a shed or in a house, not in the open air. "The weaver," wrote Barlow, "sits with his right leg bent under him upon a piece of board or mat, placed close to the edge of the pit, and depresses the treadles alternately with the great toe of the left foot." Nothing if not precise, Barlow described the pit as measuring about three feet in length by two feet in width and one and a half feet in depth. The shuttle was a boat shuttle made from the light wood of the betel-nut tree, tipped with iron points and weighing about two ounces. The loom, obviously more developed than Mill's example, also contained a suspended reed-batten (with some 2,800 dents in its forty-inch length) and revolving warp beam. Each beam was locked in place by a stick, one end of which passed through a mortice in the end of the beam, while the other end was braced against the ground. Crude but effective. The loom probably resembled the four-post loom shown in fig. 6-8.

Other even more primitive versions of the pit loom undoubtedly existed in India, and it is probable that both descriptions are accurate for different parts of the country or for different periods in history. Because so little was needed in the way of a loom "frame," the pit loom was ideally suited to an environment that offered little wood for construction. What continues to astonish textile historians is the delicacy of the fabric that was produced on such a crude tool. One frequently quoted story tells of the cow that ate an entire sari that was stretched out on the grass to dry. Another tells of the Emperor Aurungzeb, who rebuked his daughter for appearing naked when she was actually wearing seven layers of cloth.

How did they achieve such fine weaving? It is worth recounting, for it is an art that the world may never see again. Mill, in a paragraph that today would be regarded as outrageously colonial if not slanderous, attributed Indian weaving skill to the nature of both the Hindu disposition and body. In the gentler half of the paragraph, which is concerned with the latter, Mill states: "The weak and delicate frame of the Hindu is accompanied with an acuteness of external sense, particularly of touch, which is altogether unrivalled; and the flexibility of his fingers is equally remarkable. The hand of the Hindu, therefore,

constitutes an organ adapted to the finest operations of the loom, in a degree which is almost or altogether peculiar to himself." In *Textiles and Ornaments of India* John Irwin reported that the introduction of the power loom in the nineteenth century threw tens of thousands of Indian weavers out of work, and thousands starved because they found that their hands were unfit for other manual labor. The Governor General, Lord Bentinck, wrote, "The bones of cotton weavers are bleaching the plains of India. . . ."

The skill inherent in the Hindu hand found expression in finely spun cotton yarn, the finest spun by women under the age of thirty, who worked early in the morning when the dew was still on the grass and the temperature was in the low eighties. (The moisture in the air contributed to the stickiness of the cotton and kept the threads from breaking.) The spindle was a ten- to fourteen-inch needle of bamboo or iron weighted at the bottom with a pea-sized ball of clay. It was spun in the hollow of a piece of shell, and, if the air were dry, the necessary moisture was obtained by spinning the yarn over a vessel of water.

The yarn at this point was still a good eight days' worth of processing away from the loom. Yarn for the warp was first steeped in water, which was changed twice a day, for three days. On the fourth day the yarn was rinsed, reeled, and wound into skeins. The skeins were again steeped in water, then twisted tightly between two sticks, and dried in the sun. The skeins were next untwisted and placed in a mixture of water and charcoal powder, lampblack, or soot scraped from a cooking pot. They soaked in this mixture for two days and were then wrung out and hung in the shade to dry. The skeins were again reeled, steeped overnight in water, and spread out on a flat board. After being smoothed out with the hand they were sized with a rice paste mixed with a small amount of fine lime and water. After sizing the skeins were wound on large reels and spread out to dry quickly in the sun. Finally, the thread was again reeled and sorted prior to warping. It was generally divided into three grades of yarn—the finest for the right side, the next finest for the left side, and the coarsest for the center of the warp.

The yarn for the weft received a similar but abbreviated treatment, with only enough prepared for a day's work at a time. The process of preparing the weft continued on a daily basis until the cloth was completed. For a piece of cloth one yard wide by twenty yards long, this could take two men anywhere from ten days to two months, depending on the degree of fineness desired. (Although the above description refers to nineteenth-century weaving, it is believed that the same or a similar procedure was followed in ancient times.) After weaving the cloth was bleached and repaired by the dressers. The best were able to remove a broken thread twenty yards long and to *replace* it. In 1866 J. Forbes Watson wrote of these men, "Most of them are addicted to the use of opium and generally execute the

NO.1. SPINNING FINE YARN.

NO.2. WARPING.

NO.3. REELING YARN FROM A REED.

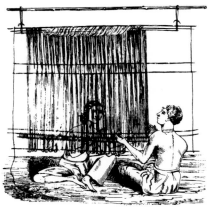

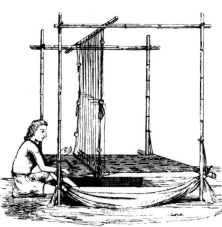

NO.4. APPLYING THE REED TO THE WARP.

NO.5. WEAVING.

NO.6. FORMING THE HEDDLES.

NO.7. STEAMING CLOTHS DURING THE PROCESS OF BLEACHING.

NO.8. ARRANGING DISPLACED THREADS IN CLOTH.

6-9: The process of spinning and weaving in India. From Watson, *The Textile Manufactures of the People of India*, 1866. Courtesy of Science and Technology Research Center, The New York Public Library, Astor, Lenox, and Tilden Foundations.

finest work whilst they are under the influence of this drug.'' The entire process, opium or no opium, was quite extraordinary (fig. 6-9).

THE COTTON ROUTE

India's monopoly on cotton textiles, at least in the Eastern Hemisphere, went virtually unchallenged for almost three thousand years. The earliest evidence of cotton weaving consists of a couple of scraps of yarn and cloth dating to c. 3000 B.C., found at Mohenjo-Daro in the Indus Valley. Frequent references to cotton in the *Sacred Institutes of Manu*, c. 800 B.C., indicate that the Hindus held the fiber in high—almost reverential—esteem. If cotton was a source of

awe to the Hindus, it was a source of bewilderment to outsiders. It baffled Herodotus in the fifth century B.C. and mystified Nearchus, Alexander's admiral, who noted as he sailed down the Indus in 327 B.C. that ''the Indians wore linen garments, the substance whereof they were made growing upon trees; and this is indeed flax, or rather something much whiter and finer than flax.''

The mystery of the cotton fiber spawned various myths concerning its origin, some of which persisted even after the truth was known. One of the more interesting is the story of the vegetable lamb, or Borametz, a miniature, fleecy animal that grew out of the stem of this mysterious plant. Friar Odoric the Bohemian, in writing of his travels in the fourteenth century, describes this strange plant-animal: "Another passing marvelous thing may be related, which, however, I saw not myself but heard from trustworthy persons. And when these [bolls] be ripe, they burst, and a little beast is found inside like a small lamb, so they have both melons [fruit of the cotton plant] and meat. And though some, peradventure, may find that hard to believe, yet it may be quite true."

Outside India cotton is first mentioned in the Annals of the Assyrian King Sennacherib (705–685 B.C.), who planted in his orchard "trees bearing wool." Whether or not Assyrian cotton survived much beyond King Sennacherib's experiments is not known, but by Hellenistic times Indian cotton was well known in Greece and Rome. It spread from India to the islands of the Persian Gulf and from there to the Arabian peninsula and probably into North Africa. Arab conquests thereafter carried it westward to Spain, where cotton cultivation is believed to have been introduced, probably at Valencia, in the ninth century. Cotton moved north slowly, not reaching England until the late thirteenth or early fourteenth century, where it was first used not for clothing but for candlewicks. Substantial trade with England did not occur until after Vasco de Gama discovered a sea route to India around the Cape of Good Hope at the end of the fifteenth century. However, cotton was not woven or spun in England until the seventeenth century when the Revocation of the Edict of Nantes (1685) sent thousands of Huguenot weavers fleeing across the Channel, a migration that signaled the beginning of a new age for England.

The pit loom may have followed some of the same trade routes as cotton, for it turns up in Sudan looking in the 1920s much like the primitive Indian pit loom. The frame, what there is of it, consisted simply of two short front posts to hold the cloth beam. The beam was prevented from rotating by a stick-and-mortice arrangement similar to the Indian loom. The warp was bunched together at the far end and tied with a rope that led away from the weaver, around a warp post stuck in the ground, and back to a post by the weaver's side where it was fastened. The batten, heddle frames, and treadles were suspended from a rod that hung from the roof beams, and the weaver worked the treadles with his feet in a pit. Because the warp was knotted together at one end, the cloth tended to narrow as it was woven. This tendency was minimized by drawing out the warp as far as possible before bunching it. In an indoor weaving shop this created quite a sight. Grace M. Crowfoot describes it: "Some of the weavers in Omdurman sit near the door of the house, so that a strong beam of light falls on their work, but the majority, especially those who work four and five in one room, their webs crossing each other till they look like a colony of giant spiders, sit in a half light which must be equally bad for their health and their craftsmanship."

Roth may have been right when he called the pit loom "rather a method of working a loom than a distinct form of loom." The loom as "method" is demonstrated in fig. 6-10, which shows a "modern" (i. e., 1922) loom from Lisht, Egypt.One expert believes that the Lisht loom was used as long ago as the Middle Kingdom in Egypt (1580–1200 B.C.), but this has not been corroborated.) Although the

6-10: Modern pit treadle loom from Lisht, Egypt. Petrie Museum, University College, London.

6-11: Egyptian warp-weighted loom. From Gilroy, *The History of Silk, Cotton, Linen, Wool and Other Fibrous Substances*, 1845.

loom is a pit treadle loom, the characteristic pit is less distinctive than the weighted warp. The loom is a curious, perhaps ingenious, hybrid. The frame is not freestanding. Two posts in the ground support the cloth beam. Unconnected to this is a separate structure, apparently embedded in the wall and floor of the house, for the sole purpose of providing a support for the suspended reed-beater. The bar that supports the heddles and pulleys is not fixed to this frame but hangs by two cords from a roof beam. The warp, after passing through the heddles, passes under a bar fixed to another separate set of implanted posts and leads upward over a final warp bar, presumably hanging from another roof beam, and down again, where the knotted warp ends are weighted with rocks. Sending the warp up instead of out, as in the Sudanese pit loom, is clearly an invention to conserve space.

A crude illustration from the mid-nineteenth century (fig. 6-11) shows the hidden superstructure, but the artist is not to be entirely trusted. For example, although the looms in this drawing have suspended, counterbalanced heddles (the pulleys and cords are clearly visible), there is no pit or any other system indicated for treadling. The missing warp in the middle of the foreground loom must reflect the artist's perception of how the warp separated when it was grouped in sections for weighting (see fig. 6-10).

A more accurate, though recent, representation is the pit loom from Syria shown in fig. 6-12. This type of loom was common in Palestine for weaving wool. A lighter version of the same loom was used for weaving cotton or a mixture of cotton and silk. Note the cord behind the weaver's shoulder for releasing additional warp as the weaving progresses. An alternate method of securing the warp ends was to weight them with stones or sandbags, as in the Lisht loom above. The warp, as it passed down and around the intermediate warp beam toward the heddles, was kept evenly spread out by nails (a kind of permanent raddle) spaced along the front of the beam. The right-hand end of the cloth beam clearly illustrates how a stick through a hole in the cloth beam kept the cloth from unrolling. The heddle frames are suspended from whippletrees, which in turn hang from a roof beam. The reed-batten, however, was not suspended from the ceiling but swung from a bar in the notched holder on top of the frame.

Whether the treadle originated in India with the pit loom is not known. Most authorities establish the invention of the treadle in China, but the weaving of both cotton and silk dates back to comparably obscure eras in India and China. The significance of the pit loom lies in the fact that it is possibly an evolutionary stage in the development of the modern treadle loom. How much it influenced subsequent loom development is unclear, but some of the trade routes from China passed through parts of northern India, and much cloth was shipped at an early period out of Indian ports. It is entirely possible that some Indian technology was passed along with the muslins.

6-12: Pit treadle loom in Majd el-Chams, Syria, common in Palestine for weaving woolen fabrics. Photograph by Shelagh Weir. Reproduced by courtesy of the Trustees of the British Museum.

Although it was used in Persia, Sudan, Egypt, and probably the Arabian peninsula, the pit treadle loom itself did not follow the course of cotton west into Europe. By the time cotton was cultivated in Europe—probably not in significant quantities until the Venetian and Milanese enterprises of the fourteenth century—another kind of loom had made its way west from the Near East. This was the horizontal treadle loom, an adaptation of a loom that is believed to have originated in China.

THE HORIZONTAL TREADLE LOOM

THE CHINESE LOOM

Weaving in China has been so closely associated with silk that one wonders what the Chinese nobility wore before discovering the secret of unraveling the silkworm's cocoon. Ancient Chinese ideograms suggest that earlier garments were made from hair or hemp, but little is known about the method of construction. The earliest Chinese loom may have resembled the backstrap loom of the Ainu, who formerly inhabited parts of coastal China, but that loom is itself a highly developed tool (see Chapter 5). Similar looms have been found in parts of modern China as far west as the Himalayas. Credit for the giant leap foward in Chinese loom technology in part belongs to the *Bombyx*

6-13: The process of silk production in China. Twelve woodblock prints by Kitagawa Utamaro. (1) Tending the newly hatched worms. Girl with feather brushes worms from paper where they have been incubating into shallow tray. After hatching, worms are fed finely chopped mulberry leaves. (2) Picking mulberry leaves. (3) Feeding the silkworms. Mulberry leaves are chopped in a shallow bucket; worms are fed in trays. As they grow, they will be moved to bamboo mats. (4) Stirring the silkworms, ending their "fourth rest." One woman brings more trays of silkworms; another carries off empty trays. (5) The "Great Awakening." Three women pick leaves from branches; one feeds worms, now voracious eaters, on bamboo mat. (6) The cocoon stage. One woman arranges cocoons in tray; another holds tray. Racks of cocoons in background. (7) The emergence of the moths. Women watching two moths laying eggs on paper. Threads attached to body keeps them on paper. Eggs are collected and placed in incubator. (8) Watching moths. (9) Winding the thread. Cocoons heated in pan of water over fire to loosen thread. (10) Stretching *mawata*. Women stretching silk floss (*mawata*) over wooden posts. Another woman hangs silk in skeins over bamboo rod. (11, 12) Silk weaving, winding off waste silk, winding bobbins. Textile Museum, Washington, D.C.

mori caterpillar. The extreme length of the silk filament—between 400 and 1,300 yards—and its inherent strength and smoothness make it ideal for warping. Because of these qualities Vivi Sylwan, an authority on ancient Chinese textiles, thinks that "the warp has been of great consequence in the construction and development of Chinese weaving methods and looms (fig. 6-13)."

The true antiquity of silk weaving in China has yet to be determined. The legend of the Princess Si-ling-chi (see Chapter 1) dates its origin to c. 2640 B.C., but scholars have placed it at widely varying dates on both sides of this figure. In Sylwan's opinion, the legend of Si-ling-chi was conceived during the era of Kung Fu-tse (Confucius) and other great philosophers, the middle of the first millenium B.C., when Chinese culture was characterized by glorification of the past. The first reliable evidence of silk weaving was discovered in remnants of silk cloth pre

4 **5** **6**

10 **11** **12**

served on Yin-period bronzes (c. 1500–1000 B.C.). The quality indicated that silk weaving was by then well developed, but how widespread it was is not known.

Nor is it known on what kind of loom silk was first warped. One possibility is the one-treadle loom, known in Japan as the *izaribata*. The *izaribata* probably derived from the simple backstrap loom, perhaps that of the Ainus. If so, and if the Ainu loom was a product of coastal China, then it is possible that a one-treadle loom also made an early appearance in China. A primitive version of the *izaribata* is illustrated in a late-nineteenth-century travel book about Korea, where silk has probably been cultivated since the second century B.C. (fig. 6-14). (The author states that cotton was also woven on this loom.) The loom frame is almost complete, except that the cloth beam, instead of being fixed to the frame, is attached to a backstrap. As on any backstrap loom, the weaver adjusts the warp tension

merely by leaning forward or backward. The broad paddles on the warp beam release one-half a revolution of warp from the beam at a time.

The operation of the heddle rod was a major technological advance (fig. 6-15). One shed was maintained by a shed rod, set between the warp beam and heddles, that separated upper and lower warp threads. The countershed was formed by tugging on a cord looped around the weaver's ankle (or possibly a toe). The opposite end of the cord, attached to a long arm, rocked a pivot bar on top of the uprights. When the pivot bar turned, the two rods attached to the weaver's side of the bar lifted the heddle rod and opened the shed.

The reed on this loom is not suspended, nor does it appear rigid enough to function as a batten. As on the Ainu loom, it probably served mainly as a warp spacer. (The boat shuttle in the weaver's right hand, however, might

6-14: Drawing of *izaribata* loom from Korea. From Cavendish, *Korea and the Sacred White Mountain*, 1894.

6-15: Shedding system on *izaribata* loom. Above: natural shed. Below: countershed. (A) Cloth beam. (B) Shed rod. (C) Heddle rod. (D) Warp beam. (E) Reed. (F) Warp depressor rod. Drawing by D. K. Burnham, University of Toronto, The Royal Ontario Museum, Art and Architecture, *Annual* 1962.

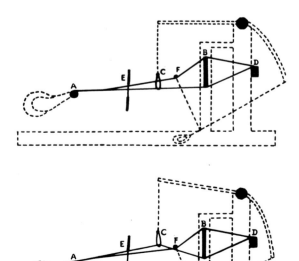

contradict this. When the reed was used as a warp spacer, the shuttle was generally batten-shaped and served the dual purposes of inserting the weft and beating it in.) Very possibly the drawing is inaccurate here.

The Japanese *izaribata* in fig. 6-16, though dated not later than the eighteenth century, represents a style of loom that must have been used in Japan for many years —exactly how many is not known. It is operated identically to the Korean loom discussed above, with a few notable differences in detail. Instead of sitting on a bench fixed to the frame the weaver sat on a rectangle of wood fastened at a slant inside the two horizontal side beams. The loop of the cord is clearly meant to go around the ankle. The warp beam is square in section, and the "paddle" ends that prevent it from revolving are not solid wood but skeletal structures. The permanent shed bar is a frame of two parallel bars and cross pieces that fits into slots in the front set of uprights.

The loom was equipped with an unusually long shuttle (approximately two feet) with a sharp lower edge. It may have been used as both boat shuttle and batten. The reed consists of slips of bamboo held between the grooves in two wooden caps. The caps are laced together with cord at the ends to hold them together, but the structure is not rigid enough to beat in the weft. The changes to a normal boat

6-16: *Izaribata* loom, Gifu Prefecture, from farm at Shirakawa village, near Hida Takayama, 18th c. Length (side beams) 53″, (cloth beams) 20½″; height (uprights) 23½″. Courtesy of the Royal Ontario Museum, Toronto.

shuttle (along with a rigid beater) and a raised, more comfortable frame were probably fairly late developments.

These later developments have not been dated precisely, although a stone relief picture has survived from the Han dynasty in China (206 B.C.–A.D. 220) that illustrates a two-treadle loom and a raised frame with seat (fig. 6-17). The picture, from one of the Wu family tombs in the province of Shan-tung, illustrates a Chinese legend from the fifth century B.C., which Sylwan relates as follows: "A woman sits at her loom turning towards a kneeling person, the shuttle she had dropped, it hangs by a thread. Twice before she has been told that her son, Tseng-tze, known for his filial piety, has committed murder, but she has not allowed it to make her anxious; the third time, however, shown here, she turns towards the kneeling person and throws down the shuttle."

Although the picture lacks detail, it does indicate that the loom was equipped with two heddle frames, one of which is apparently raised. The height of the loom and the built-in bench may mean that the Han loom was more developed than certain nineteenth-century Chinese looms. But it was not unusual in Asia for earlier, simpler weaving techniques and looms to survive alongside more highly developed forms. Fig. 6-18 illustrates what may be a more developed version of the Han loom. Some China scholars state that

the Han loom, with its warp at an oblique angle, developed from a vertical loom; others believe it evolved from the horizontal, foot-braced backstrap loom.

The Han dynasty introduced a period of vigorous expansion in China during which trade, at first primarily in silk, was established with the West. Prior to this little was known about sericulture outside China. The silkworm was first mentioned by Aristotle (384–322 B.C.) in his *De animalibus historia* as the source of silk thread, but all that the Greco-Roman world knew about silk (or *Serika*, as they called it) was that it was manufactured somewhere in the East by people whom they called *Ser*, or *Seres*, after the fabric that they made.*

There is little agreement on what kind of loom the Chinese used during the Han period to produce their figured silks. Some believe that a drawloom must have been used to produce the patterns (see Chapter 7); others believe that they could have been woven on a multiharness loom without a drawloom attachment; still others feel that the loom must have been a one-treadle loom as described above.

---

*The name "China" derived from and was introduced during the Ch'in, or T'sin, dynasty (256–206 B.C.), but it was not until the Jesuit explorations of the seventeenth century that the Chinese and Seres were identified as the same people.

6-17: Stone relief illustration of early Han treadle loom, c. 206 B.C.–A.D. 220. From Chavannes, *La Sculpture sur pierre in Chine au temps des deux dynasties Han*, 1893.

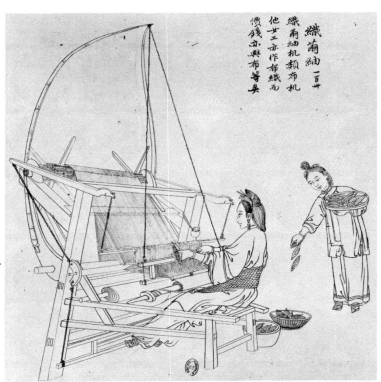

6-18: Chinese *izaribata* loom for silk and cotton. Photo: Bibliothèque Nationale, Paris.

With no surviving looms or reliable illustrations from the period, the opinions must be based on examination of Han silks. More certain is the belief that this brief period of contact with the West was instrumental in a two-way exchange of loom technologies—from Syria and Persia to the East as well as from China to the West. Although it is uncertain which specific techniques were passed to which culture, it is worth looking at a few of the possibilities.

THE MIDEAST CONNECTION

Anyone caught smuggling silkworm eggs from China suffered the death penalty. No one knows how often this punishment was imposed, but the Chinese obviously knew the value of their discovery and managed to preserve the secret of silk production for some two thousand years. The story of how knowledge of sericulture finally did reach the West ranks with the world's great adventure stories. The cast included Roman and Byzantine emperors, luxurious ladies, caravan merchants, scores of bandits, a wily princess, and a couple of clever, smuggling Nestorian monks. All this intrigue was inspired by Roman and Byzantine lust for the luxuriant silk fabric introduced to them by the Han trade routes.

The first overland silk route was created in 102 B.C. and ran from the western provinces of China along the lower rim of the Tarim Basin through Khotan and Turkestan to Bactria (Balkh) in Afghanistan (fig. 6-19). In 77 B.C. a more northerly route was established. Either was fraught with

danger. The southern route traversed the harsh wilderness of the desert in the Tarim Basin; the northern route was protected by a mountain range against the Siberian winds, but the same hills concealed tribes of marauders who looted the caravans. An enormous toll must have been taken along these routes, both in goods and in lives. At Bactria the silks were sent down to the Indian ports to the Greek and Egyptian trading agencies or on to Persia, the cities of Mesopotamia, and via Damascus to Sidon or via Jerusalem to Egypt. When the silk cloth finally reached the weaving centers of the Roman Empire, it had passed through so many hands that it was literally worth its weight in gold.

At first, as in China until about 1500 B.C., silk cloth was reserved for members of the imperial court. Weavers initially unraveled the imported silks and rewove them in their traditional patterns. They experimented with silk thread in various ways, sometimes using it as weft with a linen or woolen warp, sometimes the other way around. Silk cloth gained acceptance slowly, as it was regarded by many as too effeminate for men. The Roman Emperor Heliogabalus (A.D. 218–222), a native of Syria and the first man to wear only silk clothes, was considered to have had outrageous taste. With the change of the capital of the Roman Empire to Constantinople, however, silk grew in importance as everyday dress, and by the fourth century even commoners were wearing it.

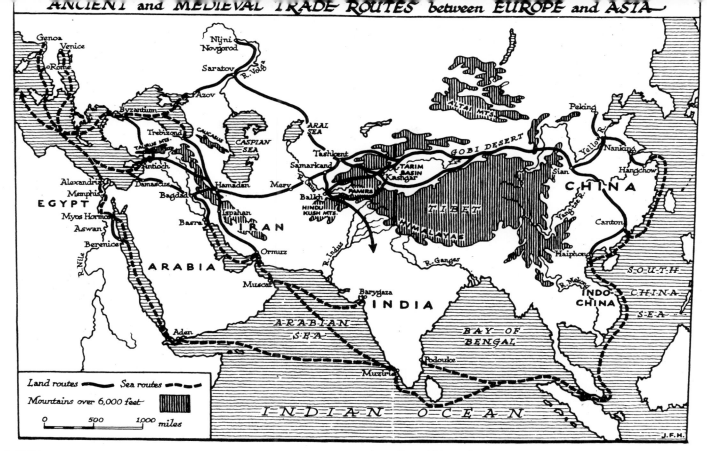

6-19: Map showing route of silk to the West. From *A History of Technology*, Vol. III, edited by Charles Singer, E. J. Holmyard and A. R. Hall. Published by Oxford University Press.

The Persian silk industry began its ascendency with the conquests of the Sassanian king Shapur II. In A.D. 355 he invaded the Roman province of Byzantium and carried back with him to Susa the best of the Syrian wool weavers. By virtue of the skill of these technicians and of Persia's pivotal geographic position, Sassanian silk soon monopolized the West. In fact, Persian dominance grew so powerful that the characteristic style of Sassanian weaving—repetition of detail face to face in mirror images—survived the Mohammedan conquests of the seventh century and persisted into the Middle Ages. Even Chinese silks reflected the Sassanian influence. In Chapter 1 we have already noted how two Nestorian monks carrying silkworm eggs in the hollows of their bamboo staves smuggled the secret of sericulture out of Khotan, a West Mongol principality in Turkestan, and into Constantinople. Their cunning enabled the Emperor Justinian to establish the basis for a silk industry in Byzantium, but it was some time before Byzantine silk encroached on the Persian monopoly.

THE WESTERN LOOM

By the time silk first arrived in the West, weaving was already a highly developed art. In Egypt linen weavers had been producing exquisite garments on both horizontal and vertical looms for centuries—most likely on looms without treadles. Although reference has been made to a treadle loom in Hellenistic Egypt in the second century B.C., no reliable evidence of such a loom has yet been discovered.

Greece and Rome knew the vertical loom—a warp-weighted loom in Greece and a two-bar loom in Rome. Again, a simple horizontal loom with treadles may have been used contemporaneously with the vertical looms, but no mention of one exists in the literature or art of the time. On the contrary, the surviving literary and artistic representations of looms all indicate a vertical loom without treadles. If a horizontal treadle loom existed at that time, it must have been used for very exclusive purposes.

As discussed in Chapter 3, most of the loom development in the Mediterranean regions was centered in the Near East. The Syrians and possibly the Palestinians were known to have been unusually experimental in the weaving of wool, much more so than, for example, the Egyptian linen weavers. The linen weavers were satisfied with their two-harness loom because tabby weave was well suited to the sleek flax fibers, which could lie close together without sticking. The wool weavers utilized the horizontal harness loom of the linen weavers, but by A.D. 256 they had developed a third harness for weaving weft twills.

As with the invention of the Chinese silk loom, the Syrian improvement stemmed from the weaving characteristics of the fiber. Weaving wool tabby was difficult because the warp yarns tended to stick together as the sheds were opened and the weaver could not throw the shuttle from selvage to selvage. Yet separating the warps only weakened the cloth. The answer, a revolutionary development in loom technology, was the addition of a third harness. Setting up the loom for twill spaced the warps in each shed far enough apart so that they did not stick. And the wool weaver, for the first time on a horizontal loom, could open a shed and throw the shuttle the entire width of the cloth.

Looms were previously designed only to accommodate the local fiber for weaving. If a new fiber were introduced, changes were made in the looms only as required to accommodate that new fiber. As a rule change came slowly, with technology lagging behind fashion. But the Near Eastern wool weavers continued to experiment, and these were the weavers that Shapur II captured and brought home to Persia.

Their two- and three-harness looms probably had no treadles at the time that silk made its way west. The silk fiber must have given Near Eastern weavers much to think about. Whereas they had previously woven twenty to thirty woolen warps per inch, with silk they faced the challenge of coping with three hundred to four hundred threads in the same space. They initially tried weaving their traditional patterns in silk on the wool looms, using techniques ap-

propriate for wool. They spun the silk, a skill unknown in the Far East, and wove weft twills without any knowledge of Chinese techniques. (The Chinese had been weaving warp twills. It is presumed that weft twills and the famed Chinese *k'o-ssŭ* tapestry techniques originated with the Syrian tapestry weavers who worked in Persia.) The experimentation of the Syrian weavers with the new fiber must have inspired changes in their loom to make it more suitable for weaving silk.

The Chinese loom for weaving figured silks, whether a drawloom or otherwise, was developed far beyond what existed in the Near East or Persia at the time of China's contact with the West. One example in the treasury of Sens Cathedral, known as the Maenad silk, indicates 1,200 to 1,400 different sheds in a fragment about five and one-half inches long. (Although this fragment probably dates from the fifth century A.D., the capabilities to produce it must have existed much earlier.) It is unlikely that a loom as sophisticated as that used for figured silks could have superseded the simpler looms used for wool and linen weaving in the Near East. The silk loom was too advanced for the wool and linen weavers to assimilate on contact. The invention of the treadle might have passed from China (or India) into the Middle and Near East during this period

6-20: West African narrow-band loom, Adamawa region, 1911. From Frobenius, *Das Sterbende Afrika*, 1923.

of contact, but there is simply not enough evidence to indicate the extent of loom technology that might have been assimilated. Most authorities agree, however, that the horizontal treadle loom that eventually reached Europe during the Middle Ages probably did not descend directly from the Chinese loom but evolved from the horizontal loom of the Syrians, which had been adapted for silk.

THE NARROW-BAND TREADLE LOOM

The African treadle loom is characterized more than anything else by the narrow width of cloth that is woven on it (fig. 6-20). It is found primarily in West Africa and is used for weaving strips of cotton cloth generally between two and six inches wide. The length of the strips might vary from a few yards up to several hundred. After the weaving was finished, the strips were cut into segments and sewn edge to edge to make a wider fabric, which was then tailored into clothes.

While looms elsewhere generally developed heavier and increasingly more sophisticated frames to support more and more moving parts as they evolved, in West Africa just the opposite seems to have occurred. The loom frame was stripped down to the barest essentials. Roth believed that the narrow-band loom may have represented various degenerate forms of more developed looms introduced into Africa, possibly by the Portuguese in the sixteenth century. Other authorities believe that the crudity of the African treadle loom indicates that it must have been of native origin. Still others have stated that the treadle loom probably came into Africa in the eleventh century with Islamic culture and crossed into West Africa from the Maghreb via one of the trans-Saharan trade routes. In her book *West African Weaving* Venice Lamb expresses the view that the loom may have originated in the Nile Valley or some other region to the east and spread southward and westward along with the cultivation of cotton. The question of its true origin, as Lamb acknowledges, has yet to be settled.

Lamb believes the narrow-band loom appeared in West Africa well before the tenth century A.D. Its use along the Senegal coast had been documented in 1455 by this report of the Portuguese explorer Cado Mosto: "They weave pieces of cotton a span wide, but never any wider, not having the art of making larger looms; so that they are obliged to sew five or six of these pieces together when they make any large work." "Not having the art of making larger looms" does not explain the longevity of the narrow-band loom in West Africa. The custom of weaving narrow bands and sewing the strips together continues to this day, often with startling and occasionally brilliant results (fig. 6-21). The assumption that, had he a wider loom, the African weaver would have abandoned the narrow-band loom, cannot be supported, particularly in fifteenth-century West Africa where he had been compet-

6-21: *Above:* Men's weave, Akan-Ashanti, Ghana, 72″ x 60″. A cloth combining 15 strips, each of a different pattern, for a black-and-white cotton robe. UCLA Museum of Cultural History. Photograph by Robert Woolard. *Below:* Men's weave, Akan-Ashanti, Ghana, 96″ x 48″. A *kente* cloth. Courtesy of The Brooklyn Museum.

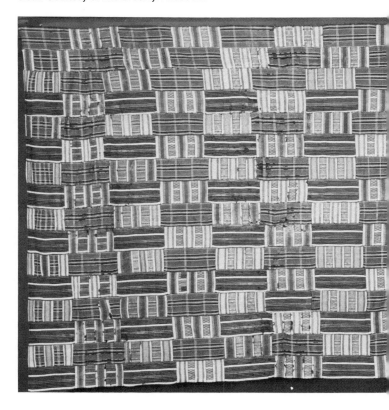

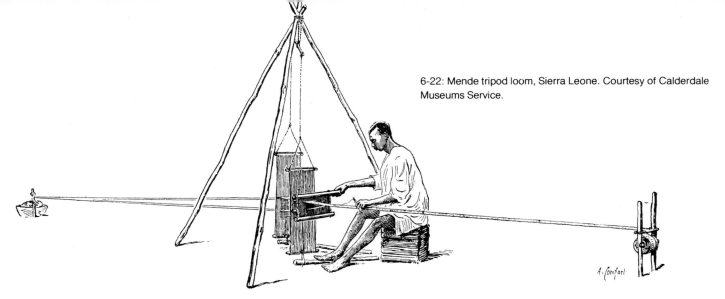

6-22: Mende tripod loom, Sierra Leone. Courtesy of Calderdale Museums Service.

ing successfully with imported clothes for several centuries. According to Lamb, the narrow-band loom has been used for weaving "bags, sheets, rugs, tent dividers, pillows, umbrellas, shawls, head cloths and pads, girdles, trousers, underwear, riding gowns, robes, smocks, hats, slung chair seating, palanquin covers, and even handkerchiefs." Several explanations for the survival of the narrow-band loom can be imagined, all of them more probable than the West African's ignorance of or inability to construct a wider loom. These would include the natural conservatism of tribal society in West Africa, the portability and ease of storage of the loom, the relative ease of weaving strong cloth in narrow strips, the fundamental changes in design that a change in width would require, and perhaps the low cost of material to start weaving on such a loom.

Two general types of narrow-band treadle looms have been used in West Africa. The tripod-frame loom of the Mende people in Sierra Leone and the Susu in Guinea is the simpler of the two (fig. 6-22). The heddle harnesses and treadles, suspended from a whippletree, hung from a tripod that could be shifted along the warp as weaving progressed. The reed-beater on the Mende loom was not suspended along with the harnesses but, as the illustration shows, was hand-held around the extended top bar of the reed. The cloth beam, secured behind the weaver by posts in the ground, rolled up the cloth as it was woven. A basket at the opposite end contained the excess warp until needed. When the weaver, with his tripod apparatus, approached the warp post, the woven cloth was presumably wound up on the cloth beam, new warp released from the basket, and the cycle repeated until the entire warp was woven. A similar loom has been used in Java and other parts of Indonesia (fig. 6-23). Instead of a tripod, however, the shed-making apparatus traveled along an overhead bar that was supported by uprights planted in the ground outside the loom bars.

With the second type of narrow-band loom the weaver sits behind a revolving cloth beam (fig. 6-24). The construction of the frame of this loom varies from one part of

6-23: Treadle loom used in Java and other places in Indonesia. Men weave on this loom. The whole harness is shifted along the bamboo rod as weaving progresses. From Jasper, *De Inlandsche Kunstnijverheid in Nederlandsch Indie*, 1912.

6-24: Two types of West African narrow-band looms. *Left:* Bariba tribesman, Northern Dahomey. The loom frame consists of six uprights pounded into the ground—two each for the cloth and warp-diverting beams and two for the bar from which the harnesses are suspended. *Below:* The loom frame consists of a tripod arrangement that contains the shedding apparatus. The cloth beam is fastened to two uprights behind the weaver. In both cases the warp is stretched out and held taut by weights on the ground. Photographs by René Gardi.

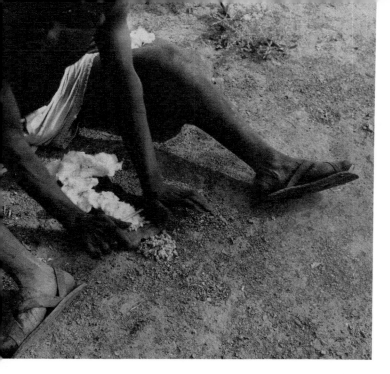

6-25: West African woman ginning cotton. Photograph by René Gardi.

6-26: West African woman spinning cotton. Photograph by René Gardi.

6-27: A weaving "shop" in Bamako, Mali. More than 100 weavers are at work here, each revealed by the bundle of warp attached to the dragstone. The weavers sit in the shade under their sheds, and only the warps can be seen in the sunlight. Courtesy of Venice and Alastair Lamb.

West Africa to another. In some places carpenters erect a rigid rectangular frame; in other places—particularly where wood is scarce—the harnesses and reed-beater might be suspended from an overhanging branch of a tree or in some other arrangement. The cloth beam of the first loom shown in fig. 6-24 is held by notched posts driven into the ground alongside the weaver. In fig. 6-20 the cloth beam floats in front of the weaver, held by ropes staked out behind the weaver and by the tension of the warp in front. In both cases the warp stretches over a diverting bar and extends across the compound to a weight or dragstone of sun-dried mud where it is anchored. As the cloth is woven and rolled up on the cloth beam, the dragstone and warp are hauled closer to the weaver.

The reed-beater, unlike that on the Mende loom, hangs from a top bar, and the harnesses hang from a simple wooden pulley, often an empty cotton-thread spool set in a carved wooden block. The treadles might be sticks, as indicated here; pieces of gourd or bone gripped between the toes; or loops of string in which the toes are inserted. The usual Ashanti loom has four harnesses—one pair for plain weave and another pair for the patterns. If pattern harnesses were not used (as in the above illustrations), the pattern wefts were darned in with the fingers.

The narrow-band loom, with local variations, is still used in West Africa today. The cotton is grown locally and ginned by rolling an iron rod over it to squeeze out the seeds (fig. 6-25). It is then fluffed up by bowing and spun on a hand spindle (fig. 6-26). The warp is prepared by winding the yarn around stakes in the ground, often leading the yarn completely around several houses in the village compound. Some of the warps extend four hundred yards.

In Mali and parts of western Nigeria the weavers sit under sheds that protect them from the sun, with their warps extending out into the compound. One shed might protect anywhere from two to thirty-five weavers (fig. 6-27). At the end of a day of weaving the looms are dismantled and taken inside for the night. One authority stated that one weaver could weave from dawn to dusk about three yards a day. In four and a half days he could weave enough strips to sew together a woman's wrapper. In West Africa only men wove on the horizontal loom. Women spun the yarn and wove on the vertical loom. It seems fairly typical of most societies that, when weaving moved out of the household and into the marketplace, men took over at the looms.

As crude or as flimsy as it might seem to western weavers, the African treadle loom commanded considerable respect in its own milieu. Among the Ashanti it has been said that old looms must not be broken up or used for purposes other than weaving. The entire loom, if it was necessary to dispose of it, was thrown into a river. The Ashanti regarded the loom as a household deity, a protector of the home. If adultery had been committed in the weaver's home, a sheep was sacrificed to the loom and to the chair of the ancestors.

The loom has traveled a long way in this chapter, from Stone Age China to present-day Nigeria. The gaps in information, particularly regarding when and how the invention of the treadle might have crossed from East to West, are profound. The discovery of silk, the invention of the treadle loom, the dissemination of the knowledge of sericulture, the cotton pit loom, the Mohammedan conquests and the spread of loom technology—the sweep is enormous. What remains unknown is perhaps even more significant than what is known, and there is no doubt that, as new evidence is unearthed, the story will change.

7. The Drawloom

The invention of [the drawloom] *was . . . as great an event in the development of the weaver's art as the printing press in the development of the printer's art.*
—J. F. Flanagan

The Chinese word for loom, chi, *implies that it is the machine* par excellence.—Joseph Needham

The drawloom represents the epitome of man's ingenuity in mastering a tool in pursuit of an art. Prior to its invention free-figured weaving—or allover pattern weaving, as it was also called—was a tedious, time-consuming process. To produce a free-figured design, a weaver had to have the means to lift *individual* warp threads or any combination of warp threads at will. (A free-figured design could also be achieved by the tapestry technique, in which individual *weft* threads were woven partway across the warp [not selvage to selvage] to complete part of a design. But the discontinuous weft of the tapestry technique produced structural weaknesses in the fabric that made the technique unsuitable for most wearing apparel. Drawloom and tapestry weaving are so dissimilar, both in technique and function, that they cannot be compared.)

One method of manipulating individual warp threads required the use of pattern sticks, which the weaver darned into the warp before beginning to weave (see fig. 5-9). A single pattern repeat might use as many as forty or more sticks, depending on the size of the pattern. Each stick functioned as a pilot for the next insertion of the shed rod. After each shot of the weft the stick closest to the weaver would be removed, permitting the warps separated by the next stick to be raised. If the pattern were to be repeated, each stick, as it was removed, had to be carefully reinserted in the same way behind the last stick. One can imagine the patience that this must have required of the weaver. In addition to Peru the pattern-stick technique has been found in Southeast Asia and adjacent areas to the southeast, and some authorities believe that it was the true precursor of the drawloom.

The invention of the drawloom itself has been variously ascribed to China, Persia, Syria, and Egypt, with dates of the earliest drawloom fabrics ranging from 400 B.C. in the State of Chhu in China to A.D. 520 in Persia. Most experts favor a Chinese provenance, though evidence from fabric analysis supports a claim for independent invention in Syria.*

The drawloom was the answer to the weaver's search for a means of weaving complex patterns that exceeded the capabilities of multiple harnesses. The number of harnesses that could hang in a loom was limited by lack of space. In general, weaving with more than twenty to twenty-four harnesses was a cumbersome task, but, according to Barlow, if the harnesses were especially thin and crowded together by staggering them vertically (fig. 7-1), as many as eighty to ninety might be used. This may sound like a prodigious number, but figured weaving by harnesses alone might require from three hundred to nine hundred or more—clearly an impossible arrangement. Each harness or combination of harnesses lifted a set of warp threads that contributed to the development of the pattern. If the warp were a fine silk of perhaps four hundred to six hundred threads per inch, the size of the figure that a harness loom could weave was necessarily quite small, even if each heddle eye contained several warps.

THE COMPOUND-HARNESS LOOM

A major advance beyond the multiple-harness loom in pattern weaving occurred with the development of the compound-harness loom, also called the shaft drawloom. (The drawloom may have preceded it historically; the evidence is inconclusive.) The compound-harness loom employed two (sometimes three) sets of harnesses: one (or two) for the ground weave and another for the figure, or pattern. The two sets of harnesses were used in tandem, the individual warps passing through heddle eyes in each set.

---

*To trace the various arguments regarding the origin of the drawloom requires a detailed discussion of weaving patterns and fabric analysis beyond the scope of this book. The interested reader should consult Sylwan (1949), Bellinger (1950-52), Forbes (1956), Simmons (1956), and Wulff (1966).

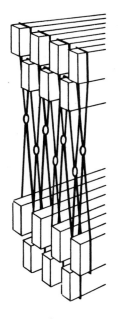

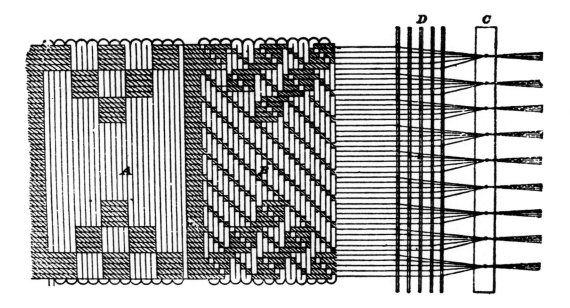

7-1: Heddles staggered vertically in the loom to conserve space. Drawing by Andy M. A. Chowanetz.

7-2: Top view of compound harness weave and harnesses. (A) Cloth woven using only figure harnesses. (B) Cloth woven using both figure and ground-weave harnesses. (C) Figure heddles. (D) Ground-weave heddles. From Barlow, *The History and Principles of Weaving by Hand and by Power,* 1878.

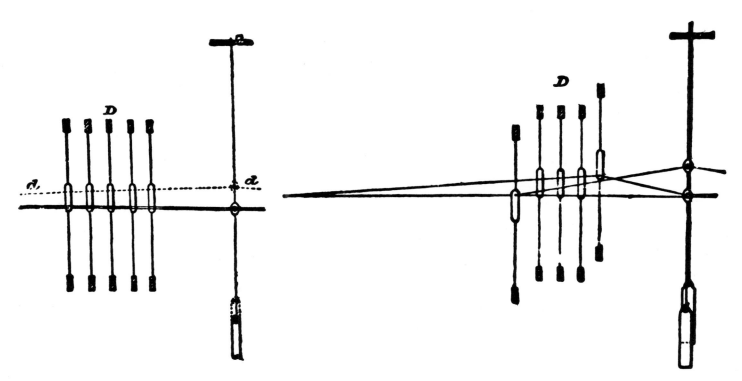

7-3: Side view of heddles for compound-harness weaving, showing elongated heddle eyes (D). After Barlow, *ibid,* 1878.

7-4: Side view of heddles for compound-harness weaving, showing how figure and ground-weave heddles can be operated independently. After Barlow, *ibid,* 1878.

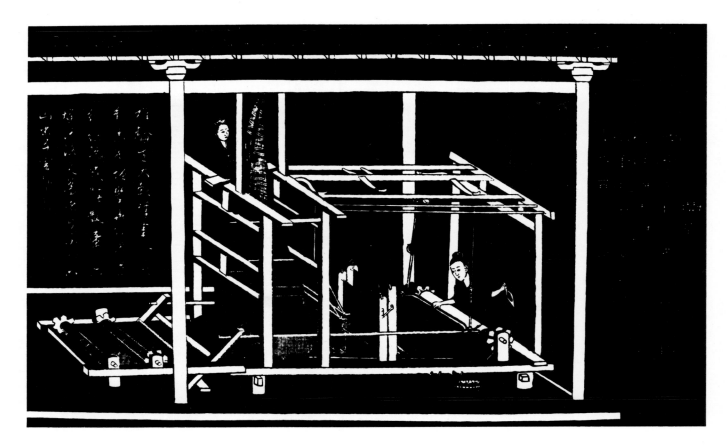

7-5: Chinese drawloom for pattern weaving. From *Keng tche t'ou*, 12th c. From Pelliot, *A Propos du Keng Tche T'ou*, 1913.

The operation of the compound harness is diagrammed in figs. 7-2, 7-3, and 7-4. Fig. 7-2 shows a top view of the compound-harness arrangement, with D representing the ground-weave harnesses and C the figure harnesses. In this example five warps pass through each heddle eye in the figure harness, then separate to pass one at a time through the ground-weave heddles. The heddle eyes of D are elongated (fig. 7-3) to permit the figure heddles (C) to be raised without interference. If they were not elongated, they would prevent the figure harnesses from operating independently. If only the figure harnesses were used, they would lift the warp in groups of five and form the pattern illustrated as A (fig. 7-2). (Note that five picks of the weft intersect with each shed formed in A.) In a like manner the ground-weave harnesses (D) can be operated independently of the figure harnesses, as illustrated at *n* (fig. 7-4). If the heddle (C) is raised, lifting five warp threads, a range of effects may be produced by varied treadling of the ground-weave harnesses. For example, *w* illustrates a raised figure heddle, but heddle *c* in the ground-weave harnesses holds down one thread (at *e*), leaving four up. Both rising and sinking harnesses are used to produce the twill shown at B (fig. 7-2).

Some claim that this is the perfect form of pattern weaving because the design is part of the very texture of the fabric and cannot be separated from it—as it could in a brocade, for example. The technique is known as *damask* weaving, a general term with over seventy-five technical definitions, named for Damascus, where some believe it originated. That claim, however, cannot be substantiated because of a lack of consensus both on a precise definition of *damask* and on the variety of fabrics woven in that ancient city.

The compound-harness loom makes it possible to weave an enormous range of small patterns into a ground weave such as a silk satin. In weaving silk it was not uncommon to thread as many as twenty or more warp threads through each eye of the figure harness in order to increase the size of the figure produced. The limitation of this kind of loom, like any harness loom, lay in the number of harnesses that the loom could accommodate. While it gave the weaver the capability of producing a tremendous variety of patterns, it did not permit the free-figured weaving of large patterns that demanded control of each individual warp thread. That extra measure of control—the mechanical repetition of any pattern, geometric or free-figured—was what the drawloom provided.

THE DRAWLOOM

THE EASTERN DRAWLOOM

The drawloom was not the first loom to require a weaver's assistant, but it was certainly the first loom to require that

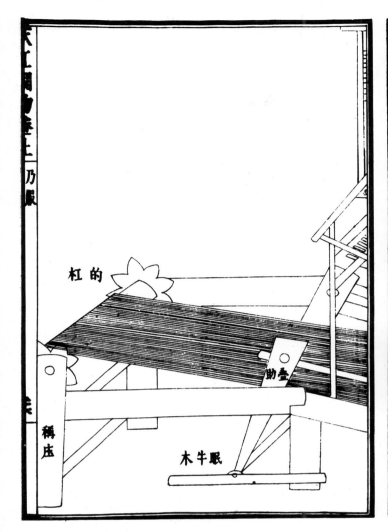

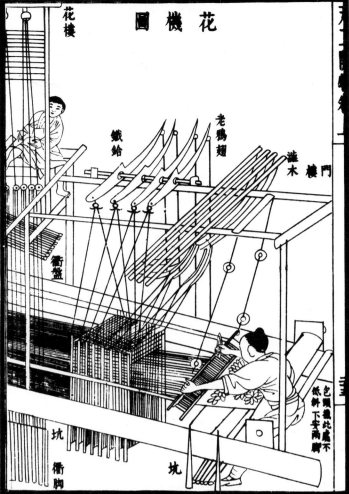

的杠

助登

稱庄

木牛眠

花機圖

花樓

鐵鈴

老鴉翅

過木樓門

衝盤

坑
衝脚

坑

包順過此處不
低斜下墊兩脚

7-6: Chinese drawloom. From *T'ien-kung K'ai-wu*, 1637.

the assistant sit perched on top of the harnesses (figs. 7-5 and 7-6). From that vantage the drawboy (sometimes drawgirl), as the assistant was called, lifted the figure heddles in the predetermined order necessary to form the desired pattern. His only problem lay in knowing which drawstring to lift when. As drawlooms evolved, various techniques were developed to organize the order of lifting the drawstrings, a process the Chinese called—simply and aptly—*pang hua*, "pulling the flowers."

In Persia the drawloom still in use in the mid-1960s (fig. 7-7) probably differed little from those developed during Sassanian times. The warp, weighted with sandbags, was divided into upper and lower warps that passed under rods pegged to the floor and then over pulleys suspended from a high back beam. The drawboy sat on a platform above the warp, with the drawstrings for the figure harnesses hanging just in front of him (fig. 7-8). The drawstrings were fastened at the top to a wooden support near the ceiling. Just above the warp the drawstrings were tied to a horizontal gut string that traversed the width of the warp (fig. 7-9), then continued beneath the warp, where they

were weighted to pull the figure heddles back down when the drawstring was released. (Note the pit in fig. 7-5, used for the purpose of containing the weights.) As the diagram shows, each of the horizontal strings in the cross harness lifted a number of warp threads, the number depending on how many times the pattern was to be repeated. Thus, when one drawstring was pulled, four warps (in the example in fig. 7-9) were lifted for a pattern with four repeats across. The cross harness reduced by four times the number of vertical drawstrings that the drawboy had to manipulate.

On the Persian drawloom the sequence of drawstrings was organized for the drawboy when the weaver threaded the heddles for the design. All the vertical drawstrings that formed the first shed of the figure were encircled with a loop of string, which was then carefully hung to one side. The drawstrings for the second shed were similarly looped together, and the loop hung next to the first. The drawstrings continued to be looped and the loops placed in sequence until all the drawstrings that formed the complete figure had been organized for pulling in a neat row of loops. As the weaving progressed, the drawboy merely removed the loops in order, pulled them at the weaver's command,

7-8: Drawboy pulling harnesses atop the Persian drawloom. Reprinted from *The Traditional Crafts of Persia* by Hans. E. Wulff by permission of The M.I.T. Press, Cambridge, Massachusetts. Copyright © 1966 by The Massachusetts Institute of Technology.

7-7: Persian drawloom. Reprinted from *The Traditional Crafts of Persia* by Hans E. Wulff by permission of The M.I.T. Press, Cambridge, Massachusetts. Copyright © 1966 by The Massachusetts Institute of Technology.

7-9: Simplified diagram of Persian cross harness. Drawing by Andy M. A. Chowanetz.

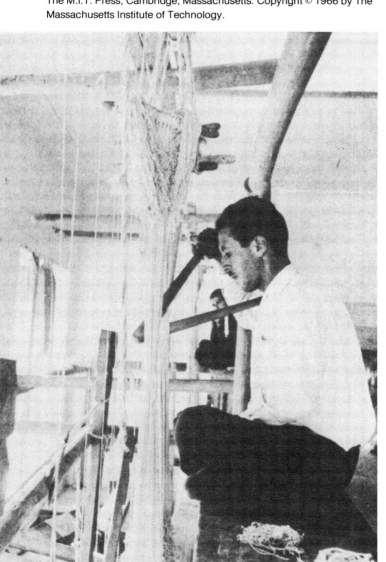

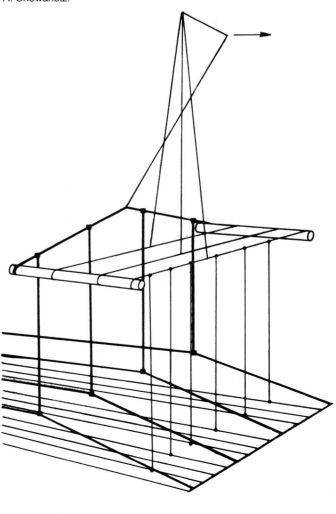

and placed them carefully in order over another rod. In between each figure weft the weaver inserted a ground or binder weft in a shed opened with foot treadles. When all the loops had been placed from one side to the other, the figure was complete. If the design called for a pattern repeat, the drawboy simply repeated the process in the same order (1,2,3,4,5, etc; 1,2,3,4,5, etc.). Or if a mirror image was desired, he pulled the loops in reverse order (e.g., 1,2,3,4,5,4,3,2,1). In China the drawboy followed a printed or written draft of the pattern, but other methods were also used. Sylwan speculates that the drawboy might have sung the pattern, a technique reported in modern India. Or perhaps the drawstrings were numbered or marked in some other way to indicate the pattern. Possibly the weaver simply told the drawboy the order.

The *T'ien-Kung K'ai-Wu,* a seventeenth-century treatise on Chinese technology by Sung Ying-Hsing, suggests that the design of a cloth was made by an artist other than the weaver, who may not have known what the pattern would be until after it was woven. The artists painted the fabric design in color on paper and translated it precisely—to the thousandth of an inch—into the silk threads used for the weaving pattern. The pattern guided the drawboy in lifting the heddles. Even if the weaver did not know what the fabric pattern and color would be, he merely followed the specifications of the pattern and watched the figure appear as he wove.

Various illustrations, most of them from the seventeenth

7-10: Chinese drawloom. Courtesy of Victoria and Albert Museum.

129

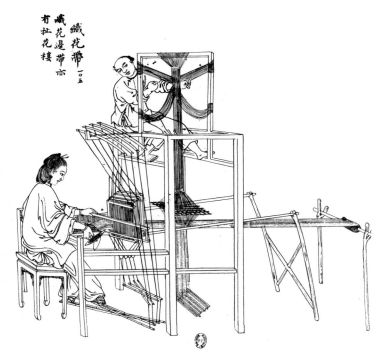

織花帶一○五
藏花邊帶布
有扯花樓

7-11: Chinese drawloom. Photo: Bibliothèque Nationale, Paris.

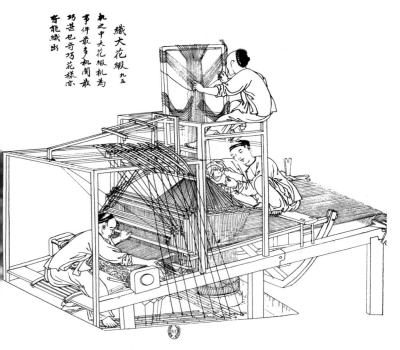

織大花緞九五
九之中大花緞机為
事件最多机閣最
巧甚巴奇巧花樣亦
皆能織出

7-12: Chinese drawloom for silk and gold brocade. Photo: Bibliothèque Nationale, Paris.

century or later, depict Chinese drawlooms. Fig. 7-10, probably the most frequently reproduced, may be among the least accurate renderings in detail. It differs from the Persian drawloom in several important respects. It shows no cross harness for lifting with one drawstring the several warp threads that repeat the same part of each pattern across the warp. Perhaps each cord that rests over the drawgirl's right shoulder was looped around all the drawstrings that had to be raised for a single pattern shed.

Neither this illustration nor fig. 7-5 makes this detail clear. (Note how the loom frame in figs. 7-6, 7-10, and 7-12 slopes toward the weaver to add force to the pull of the beater.)

Figs. 7-5, 7-6, and 7-10 all illustrate compound-harness drawlooms. In fig. 7-10, for example, one set of five harnesses, controlled by gibbet levers, would be used for treadling, for example, a satin ground weave. The second set of eight harnesses might change the ground weave or alter the design. These harnesses form a sinking, not a rising, shed. When the treadle is released, the harness is returned to its place by the elasticity in the supporting rods overhead. Figs. 7-11 and 7-12 illustrate more clearly how the harnesses may have been supported and tied to the treadles. (In fig. 7-10 the weaver is about to insert the shuttle into a shed that is formed by neither set of harnesses.) Figs. 7-11 and 7-12 illustrate a kind of grid arrangement along the vertical drawstrings, which must have functioned similarly to the cross-harness arrangement in the Persian drawloom. On the European drawloom this device was known as the *comber board*. How it operated is described below.

THE EUROPEAN DRAWLOOM
Not much is known about the European drawloom prior to the seventeenth century. No medieval illustrations have been found, and we shall see that there are good reasons to suppose that the improved drawloom was a later development.

One significant difference in the improved European drawloom was a simple invention that permitted the drawboy to work the figure drawstrings from the side of the loom instead of from the top. The diagram in fig. 7-14 illustrates how the figure harness was manipulated. D represents the ground-weave harnesses, which are worked by foot treadles. At C the figure heddles are distributed across the width of the warp by the comber board. Each heddle is weighted with a thin strip of lead wire, called a *lingoe*, which returns the heddle to its proper place after it is raised. (The heddle on the drawloom actually consists of three separate parts: at the center is the eye, or *mail*, through which the warp is threaded; above and below the eye are fastened the *couplings*, which tie to the neck cords above and to the lingoes below.) Each lingoe might weigh no more than an ounce, but for a simple silk pattern the drawboy might have to pull, including the friction of the cords, some thirty-six pounds—and hold it while the ground is woven. Various devices with a mechanical advantage were added to assist him. One example, a drawing fork, is illustrated in fig. 7-13.

The cords from each figure heddle converge below the pulley box (P), forming what is called the *neck*. (See fig. 7-15 for a clearer illustration.) The cords that will be raised together are attached to a single cord that enters the pulley

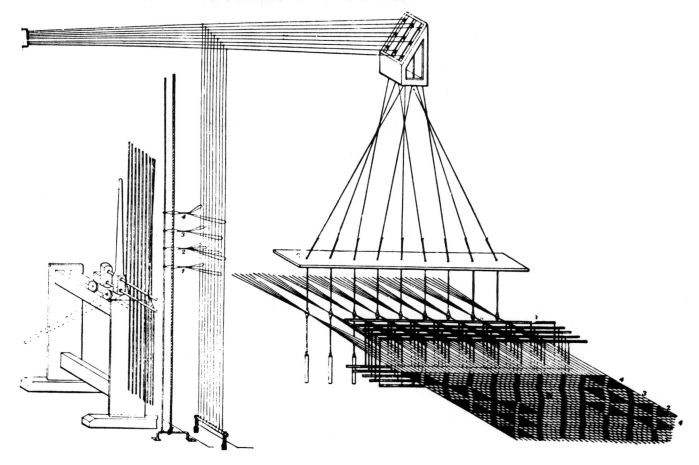

7-13: Diagram of drawloom monture and comber board. From Barlow, op. cit., 1878.

7-14: Drawing fork to assist drawboy in raising harnesses. From Barlow, 1878.

7-15: Detail of comber board and bottom of pulley box. At left, detail of the heddle and eye. From Barlow, 1878.

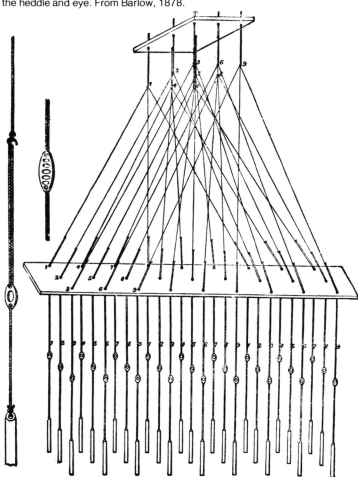

box from the bottom. This cord passes over a pulley and exits horizontally as a *tail cord* (T). The tail cords are then fastened to a wall or some other permanent fixture (L).

Attached to the tail cords and running down to a rod at the floor were a series of vertical drawstrings called *simples* (S). It is apparent that, if any of these cords were pulled, they in turn pulled the tail cords that ran through the pulley box and lifted the individual heddles accordingly. A drawboy could stand at the side of the loom and do the same work that was previously done from above. To aid him in pulling the drawstrings in proper order, loops (or *lashes*) were attached to the simples and organized along the heavier vertical cords, called *guides*, shown to the left of the simples.

If there is one item of singular importance in this arrangement, it is the comber board (fig. 7-15).* The comber board, a refinement of the Persian cross harness, keeps the heddles in place and in a relatively small space does the work of an utterly impossible number of heddle harnesses. (Note in fig. 7-15 that the holes in the comber board are drilled on the diagonal so that the eyes of one row of heddles do not obstruct the eyes in the next row.)

*According to Luther Hooper, the comber board originally was called a *camber* board, after *cambers*, the early name for lateral design repeats.

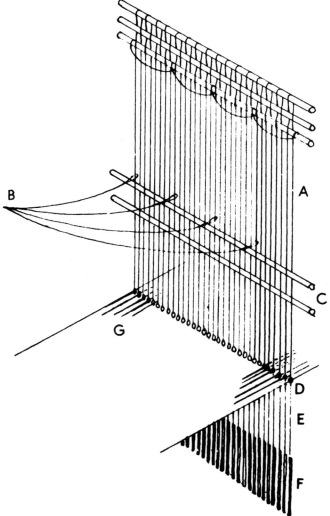

7-16: Shroud of St. Josse in weft-faced figured twill weave. Islamic, 10th c. The inscription below does not reverse with the rest of the design. It reads: "The Glory and Prosperity of the Captain, Abū Mansūr the Mighty, may God lengthen [his days]." Islamic Section, Musée du Louvre, Paris.

7-17: Probable arrangement of the early European figure harness without neck. (A, E) Harness cords. (B) Drawcords: one cord to each harness figure or repeat. (C) Cross sticks to keep the harness cords in correct order. (D) Harness mails (eyes). (F) Wire weights (lingoes) to keep the harness cords taut. (G) Warp. From *A History of Technology*, Vol. III, edited by Charles Singer, E. J. Holmyard, and A. R. Hall. Published by Oxford University Press.

The diagram shows a comber board perforated with 27 holes for 27 heddles, but this is inadequate to illustrate the advantage of the comber board. Hooper tells of a linen tablecloth woven in the mid-nineteenth century on a loom with a comber board drilled for 4,200 heddles, each one under separate control.

Evidence that the comber board was not a feature of the early medieval loom is to be found in the design of certain medieval silks. A number of silks have been found with the figure woven by point-repeat (a mirror image of the figure, the "point" being the point where the reversal begins and ends), but some of these silks also include a portion of the design, such as the inscription in fig. 7-16, which does *not* reverse. A drawloom with a comber board and neck cords would necessarily reverse *all* elements of the design. The looms that wove silks such as that shown in fig. 7-16 must

7-18: Model of loom invented by Claude Dangon about 1605. Courtesy of Musée Historique des Tissus, Lyon.

METIER A LA GRANDE TIRE

INVENTE VERS 1605 PAR CLAUDE DANGON TISSEUR LYONNAIS, IL CONSTITUAIT
UN PROGRES SENSIBLE SUR LES METIERS EN USAGE EN FRANCE DEPUIS LA FIN DU XVᵉ SIECLE
MAIS L'OUVRIER TISSEUR DEVAIT ETRE ASSISTE D'UN OU PLUSIEURS TIREURS DE LACS
C'EST SUR CE METIER QU'ETAIENT TISSES LES "GRANDS FACONNES" JUSQU'A LA FIN DU XVIIIᵉ SIECLE

have been the same as those used in contemporary Byzantium and therefore the same as those carried into medieval Europe. The figure harness of the early European drawloom probably resembled that shown in fig. 7-17, an arrangement by which some elements of the design could be repeated in reverse without affecting other elements.

The simples that enabled the drawboy to work alongside the loom (fig. 7-18) are believed to have been invented in about 1605 by Claude Dangon of Lyons. One limitation on the figure harness had been the weight of the number of lingoes that the drawboy had to lift. With the addition of the drawing fork, or lever (see fig. 7-14), the capacity of the figure harness was tripled. This improved drawloom, known as the *lever drawloom*, continued in use for damasks until the early nineteenth century.

Another method of raising the figure harnesses, known as the *button drawloom* (fig. 7-19), is said to have been invented in the fifteenth century by Jean le Calabrais. On this loom loops were tied around the tail cords, and those that would be pulled to open each figure shed were knotted together. A cord tied to this knot passed through holes in another board and terminated in a button that prevented the cord from slipping back through the hole. To open each figure shed, the weaver simply pulled the buttons in the prearranged order. While improved versions of the button drawloom continued in use until the end of the eighteenth century, it was impractical for large-figured fabrics because of the number of buttons required.

THE MODERN DRAWLOOM

Some experts believe that by the end of the tenth century A.D. silk weaving east of the Dardanelles—from Constantinople to India to China—had reached a peak seldom surpassed either in quality of workmanship or in beauty or ingenuity of design. The mechanism of the drawloom could perform all the functions that were ever required of it, then or since. In Hooper's opinion, all the modern improvements on the drawloom have been made on the mechanisms above the board shown at the top of fig. 7-15, ''and they only accelerate the speed of working, or affect some unessential detail of procedure. They do not touch the principles of the intersection of threads, in which the whole art and mystery of weaving consists.'' In fact, as will be seen, many textile historians believe that the impulse toward mechanization had a deleterious effect on the quality of textiles produced.

Silk weaving and the drawloom probably entered Europe through Sicily and Venice in the twelfth century with the

7-19: The button loom for weaving figured silks. Courtesy of Ciba-Geigy Ltd.

Saracens. Although the art of silk weaving may have existed in Spain as early as the ninth or tenth century, the textiles of the conquering Moors remained too Arabian in design to gain popularity in Europe, even as late as the fourteenth and fifteenth centuries. In Italy, however, eastern design was blended wtih native elements, and the combination of freshness and familiarity that resulted probably accounted for the acceptance of silk weaving there. Silk techniques may have existed in France by the mid-thirteenth century, but Italy maintained a virtual monopoly on European silk weaving until the late fifteenth century when a few Italian weavers escaped into France. There, encouraged by French officials, they began weaving silk. Others followed, and before long a French silk industry had blossomed in Lyons.

European figured weaving reached its pinnacle during the eighteenth century. Throughout this period various attempts were made to mechanize the operation of the drawloom. It is likely that pressure from the Indian silk trade intensified the thrust toward mechanization. It took a skilled weaver about two weeks just to set up the drawstrings and lashes on a simple drawloom for a single pattern. And each time the pattern was altered, he had to repeat the process. Often as many as three women, working long hours, labored as drawers on a single loom. Mechanization eventually put an end to that particular misery.

The history of mechanized looms has been covered thoroughly in other books, but the following summarizes briefly the inventions that culminated in the jacquard mechanism, the device that formed the basis for all industrial figured weaving today.

One of the earliest efforts at mechanizing the drawloom was that of Joseph Mason, an Englishman, who patented in 1687 a machine that he described as "an engine by the help of which a weaver may perform the whole work of weaving such stuffe as the greate weaving trade of Norwich doth now depend on, without the help of a draughtboy, which engine hath been tryed and found out to be of greate use to the said weaving trade."

Despite Mason's claims, his invention was not practical. The first significant contribution is generally credited to Basile Bouchon, a Frenchman, who in 1725 invented a device for selecting automatically which simples to pull (fig. 7-20). Cords of the simple were threaded through eyes in a row of needles that could slide in a box. Paper, perforated according to the desired pattern, was passed around a perforated cylinder that was pushed against the box containing the needles. Those needles that slid through the holes remained still, while the others, which hit unperforated paper, were pushed back, along with the cords attached to them. The selected cords were then pulled down by a foot-operated comb that engaged beads attached to the cords. The cylinder of paper was rotated with each pick of the shuttle, and a new set of holes selected the cords for the next pattern shed.

A few years later M. Falcon improved on Bouchon's invention by adding several rows of needles and replacing the perforated paper with perforated cards linked edge to edge. Each card represented the selection of needles for one shot of the weft. This simplified pattern changing, but the cards still had to be pressed against the needles by a hand-held perforated platen.

In 1745 Jacques de Vaucanson put the selecting box on top of the loom and removed the simples and tail cords altogether (fig. 7-21). Perforated cards passed around a sliding cylinder and selected the needles, which acted directly on hooks attached to the neck cords. The hooks were lifted by an iron bar called a *griffe*. Whether or not Vaucanson's invention worked is not known. One historian states that he stopped work on it because of hostility from the textile workers of Lyons. Another, adding detail, notes simply that Vaucanson was received in Lyons by showers of stones. This famed inventor of mechanical marvels, such as an automatic flute player, avenged himself by building an automatic weaver in the shape of an ass—and it actually worked!

The task of perfecting Vaucanson's loom fell to Joseph Marie Jacquard (1752–1834), also from Lyons, who came to the attention of the French government in Paris for inventing a machine that automatically tied knots to make fishing nets. In 1804 he produced what is commonly but misleadingly called the jacquard loom (fig. 7-22). It was actually a treadle-operated automatic shedding mechanism that could be mounted on top of any treadle loom with the frame to support it. So successful was his device that by 1812 it was fitted to 18,000 looms in Lyons. Although his mechanism brought great prosperity to Lyons, Jacquard himself, like others whose inventions had threatened the livelihood of textile workers, was persecuted and died in poverty.

His machine (fig. 7-23) employed a quadrangular "cylinder" that carried an endless chain of cards perforated according to the desired pattern. Only those needles that penetrated the cards and cylinder moved into a position for the hooks above to be lifted by the griffe. The hooks of the needles that did not penetrate were pushed out of the way. When the griffe lifted the hooks of the selected needles, a pattern shed was opened. Depressing and releasing a single treadle read the pattern card, opened the pattern shed, revolved the cylinder a quarter turn to present the next pattern card, closed the pattern shed, and aligned the card against the head in preparation for the next pattern shed. One historian states that prior to Jacquard's invention the children (or women) who operated

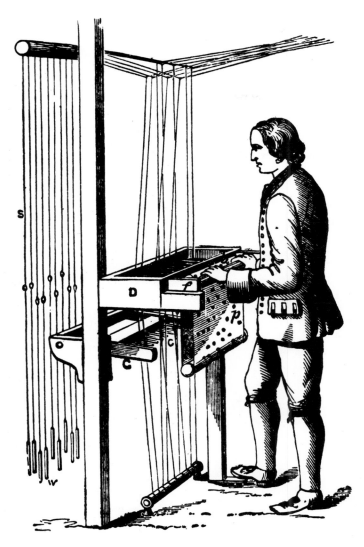

7-20: Bouchon's device for selecting simples, 1720. (S) Leashes. (W) Lingoes. (D) Needle box. (b) Cylinder. (p) Perforated paper. (g) Comb bar acting on knots or beads on vertical simples. From Barlow, op. cit., 1878.

7-21: Vaucanson's improved loom, 1750. Courtesy of Ciba-Geigy Ltd.

7-22: Model of a jacquard-type loom. Courtesy of Musée Historique des Tissus, Lyon.

the draw harnesses in textile factories sat in cramped quarters in rooms filled with floating dust and fibers and often "died before living out half their days." The jacquard machine put the drawboy out of work—perhaps to his ultimate benefit—and enabled design changes, previously a laborious, time-consuming process, to be made within an hour or so.

The ease with which patterns could be changed liberated the textile designer, and quantity began to overwhelm quality with a vengeance. The nineteenth-century market was flooded with all kinds of designs. Some textile historians believe that the mechanized drawloom was responsible for the continual quest for novelty, the constant com-

petition for the public's attention that still plagues us today.

Most of the basic figured weaves, however, were developed during the Middle Ages. Later developments merely refined or extended already established principles and streamlined production. All this, of course, refers to drawloom weaving, a special technique for figured fabrics. Most of the cloth for everyday wear, the cloth more closely associated with the increasingly specialized and closely regulated textile guilds (linens, fustians, and woolens), was woven on the simple two- (and later three- and four-) harness treadle loom. It is this loom, which appeared during the medieval era in Europe, that is the direct ancestor of the loom that most handweavers use today.

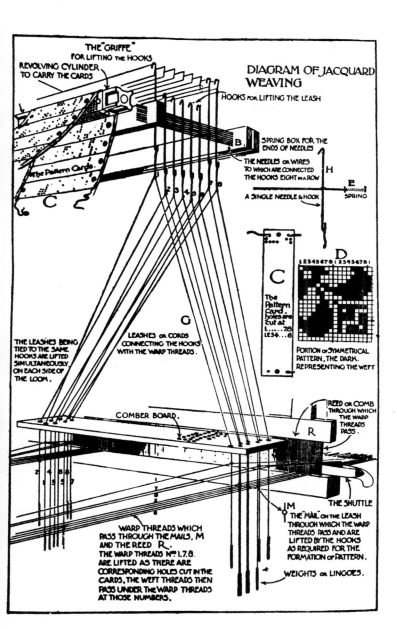

DIAGRAM OF JACQUARD WEAVING

THE "GRIFFE" FOR LIFTING THE HOOKS

REVOLVING CYLINDER TO CARRY THE CARDS

HOOKS FOR LIFTING THE LEASH

SPRING BOX FOR THE ENDS OF NEEDLES

THE NEEDLES OR WIRES TO WHICH ARE CONNECTED THE HOOKS EIGHT IN A ROW

A SINGLE NEEDLE & HOOK

The Pattern Cards

B

C

H

E SPRING

C The Pattern Card, holes are cut at 1......78, 1234...8

D

1 2 3 4 5 6 7 8 1 2 3 4 5 6 7 8

PORTION OF SYMMETRICAL PATTERN, THE DARK REPRESENTING THE WEFT

THE LEASHES BEING TIED TO THE SAME HOOKS ARE LIFTED SIMULTANEOUSLY ON EACH SIDE OF THE LOOM.

LEASHES OR CORDS CONNECTING THE HOOKS WITH THE WARP THREADS.

G

COMBER BOARD.

REED OR COMB THROUGH WHICH THE WARP THREADS PASS.

R

THE SHUTTLE

M

THE "MAIL" ON THE LEASH THROUGH WHICH THE WARP THREADS PASS AND ARE LIFTED BY THE HOOKS AS REQUIRED FOR THE FORMATION OF PATTERN.

WARP THREADS WHICH PASS THROUGH THE MAILS, M AND THE REED R. THE WARP THREADS N° 1.7.8. ARE LIFTED AS THERE ARE CORRESPONDING HOLES CUT IN THE CARDS, THE WEFT THREADS THEN PASS UNDER THE WARP THREADS AT THOSE NUMBERS.

WEIGHTS OR LINGOES.

7-23: *Above:* Diagram of working mechanism of jacquard loom. Courtesy of B. T. Batsford, Ltd. *Right:* Photograph or working mechanism of jacquard loom. National Museum of Natural History, Smithsonian Institution.

8. The Modern Loom

The medieval loom as it is reproduced by innumerable reliefs, paintings, and drawings, still persists in its principal elements in the mechanical looms of today.
—G. Schaeffer

The extraordinary fact about the history of looms is that their basic structure has not altered in five thousand years. There has been no need.—Lili Blumenau

THE HORIZONTAL LOOM

DEVELOPMENT OF THE TEXTILE INDUSTRY

Modern historians have shown us that the Middle Ages, far from being a dismal hiatus between the classical world and the Renaissance, introduced an era of great technological invention and major scientific advance. In fact, some have gone so far as to refer to a twelfth-century industrial revolution. Some of the technological innovations were applicable to the embryonic textile industry. For example, using the rediscovered principle of the water wheel, which had not been used as a source of power since Roman times, the medieval craftsman harnessed water power to activate a series of automatic hammers for the fulling of cloth, a process to soften, shrink, and partially felt the woven fabric (fig. 8-1). This had previously been accomplished by tramping on the cloth in a large vat (fig. 8-2).

With one exception medieval technology had very little to do with the development of the loom. That one exception was the introduction of the horizontal loom itself, which, as far as anyone knows, was not a western invention at all but probably an import from the Near East. It seems to have appeared, complete in all its essentials, sometime during the eleventh century. Some say about the year 1000; others say closer to mid-century. The earliest written record of such a loom that has been discovered thus far occurs in the writings of Rashi (1040–1105), "the father of Mishnah commentators." Rashi, who lived in France most of his life, wrote that men wove with their feet, while women used a cane that moved up and down.

The woman's cane refers to the sword beater that the weaver on the warp-weighted loom wielded to beat up each passage of the weft. Rashi's comment reflects not only the beginning of the change from a warp-weighted to a horizontal loom but also a change from a female to a male occupation. For a period the two kinds of looms must have existed side by side. The warp-weighted loom survived in parts of eastern Europe through the nineteenth century and is still used in Norway today.

An illustration from an early thirteenth-century pattern book (fig. 8-3) depicts the two looms during this period of change. At the left a woman is shown weaving on a warp-weighted loom. She holds a sword beater in her right hand. The twisted loops at the bottom of the warp may indicate that the warp, perhaps longer than the height of the loom, was chained to keep it off the ground. The figure to the right probably represents someone weaving on a treadle loom, but the drawing leaves much to the imagination. The figure in the lunette above, however, is clearly holding a shuttle and reed, accessories of the horizontal, not the warp-weighted, loom. Significantly, the figure at the warp-weighted loom is a woman. The others, figures associated with the horizontal loom, are all men.

The appearance of men at the looms signals the birth of European weaving as a commercial enterprise, and the credit must go to the introduction of the horizontal loom. It was now possible to weave long lengths of cloth at a speed that made the warp-weighted loom appear primitive by comparison. The textile crafts spearheaded what has been called the "commercial revolution" of the Middle Ages. The explosion of weaving into the first *grande industrie* coincided with and perhaps resulted from the growth of towns and the tremendous expansion of pan-European trade.

Cloth production was centered in Flanders and dominated the entire northwest section of northern Europe. In the twelfth and early thirteenth centuries, the cloth was sold at the various trade fairs, the best known being those at Troyes, Langres, Rheims, and Laon in Champagne. But in the late thirteenth century the Flemish weaving towns of Ghent, Ypres, and Douai, served through the port of Bruges, began to overshadow the fairs of Champagne, and by the end of the century Flemish cloth was being exported

8-1: Medieval fulling mill, powered by an undershot water wheel, 1617. From *A History of Technology*, Vol. III, edited by Charles Singer, E. J. Holmyard, and A. R. Hall. Published by Oxford University Press.

8-3: Looms shown in a pattern book, MS from the Monastery Rein (Reun) in Austria, 13th c. From Hermann, *Die deutschen romanischen Handschriften*, 1926.

8-2: Fuller trampling cloth in the vat. From the painted window of the Clothiers' Guild, Semur-en-Auxois catherdral, Côte d'Or, c. 1460. After © photo, Archives Photographiques, Paris, From *A History of Technology*, Vol. III, edited by Charles Singer, E. J. Holmyard, and A. R. Hall. Published by Oxford University Press.

all over the known world. (Over half the estimated 50,000 people in Ghent, the largest city in northern Europe, were probably engaged in the woolen industry. The proportion was even higher in Ypres. Although a smaller city, in 1313 Ypres produced 40,000 pieces of cloth, while Troyes, the capital of Champagne, barely produced 2,000 pieces a year.)

Ypres developed into an important linen center, well known for its "cloth d'Ypres," a diamond-shaped table linen from which our word *diaper* derives. Silk remained a luxury fabric that was produced in the silk centers of southern Europe at Lucca, Venice, and Genoa. Cotton, although cultivated by the Moors in Spain at least as early as the early Middle Ages, came late to the large cloth centers in the North. After a period of use mainly as candlewicks it was spun as the weft in a coarse cloth with a linen warp called *fustian*, a name derived from the ancient Egyptian town of al-Fustāt (now part of modern Cairo).

As significant as these fibers were, none of them could match the commercial importance of wool. Flanders, the nucleus of the wool trade, depended for its prosperity on access to English wool, particularly that of the superior sheep raised by the Cistercian and Premonstratensian monks. Wool ("the jewel of this realm") became so important to England's economy that the Lord High Chancellor in the House of Lords still sits upon a woolsack, symbol of the nation's former wealth. One wealthy merchant of the fourteenth century had scratched these lines in the stained-glass windows of his house for all to see:

> I thank God and ever shall
> It is the sheepe hath payed for all.

And pay for all it did. Fifty thousand sacks of wool paid for the ransom of Richard I, captured by Leopold II, margrave of Austria, as Richard returned from the Crusades. It paid for the military and political adventures of Edward I in the thirteenth century, and taxes and loans on wool by Edward III in the fourteenth century paid for the early stages of the Hundred Years' War. The merchandising of wool and wool cloth, once production had exceeded local demands, created the wealthiest men in Europe in the thirteenth century—the cloth merchants. This new bourgeoisie controlled the towns that sprouted in the areas of greatest commerce—Flanders, the Rhineland, southern and northern France, and north-central Italy.

Urbanization brought with it the development of guilds, beginning with the merchant guilds in the late eleventh and early twelfth centuries. As the industries became increasingly specialized, they soon subdivided into the various craft guilds. Specialization was an urban phenomenon. In rural areas peasants continued to spin, weave, and dye yarn as they always had—at least through the fourteenth century—until cloth became cheap enough for peasants to

consider buying it themselves. One historian has written that the best Flemish clothmakers utilized over seventy different specialists. Another has put the figure at twenty-six but either way, considering that in earlier times the weaver performed all the tasks himself, the number of specialists is surprising. A mere fourteen of them are indicated in a poem by Richard Watts: culler, dyer, oiler, mixer, stock-carder, kneecarder, spinster, weaver, brayer, burler, fuller, rower, shearman, and drawer. Much of the specialization involved processes that occurred after the weaving of the cloth. The brayer pounded and scoured the cloth to remove the oil and dirt; the burler picked out the loose threads and knots. The tramping by the fuller matted the fibers together in a kind of felting that softened the cloth and obscured the weave. The rower teased a nap up on the cloth with teasles set in a frame, and the shearman cut the nap even and smooth. The drawer mended any holes in the cloth caused by broken threads.

These finishing processes, rather than the weaving itself, distinguished the product of the horizontal loom from that of the warp-weighted loom. In addition to refining the product the finishing procedures swelled the ranks of the textile workers. A Suffolk clothier who in 1618 made twenty broadcloths a week would employ in various ways five hundred persons. While this large an operation was rare in medieval times, the textile industry was still one of great proportions.

Each specialty developed its own guild, which, as the guild grew, served several purposes. In general, it acted to preserve the status quo. It protected the monopoly of the town market against outsiders; it guaranteed full employment by restricting membership; it promoted the economic welfare of its members and regulated working hours and procedures; it established a system of craft training. At the same time its many regulations protected the consumer by assuring a uniform product at uniform prices. All first-class cloth had the same-quality weft and warp and a set number of warp threads that guaranteed the closeness of the texture. In Provins, for example, the number was 2,200. Cloth with only 2,000 warps (*vingtaines*) was considered cheap. The earliest preserved regulation from Ypres dates from 1213 and concerns the quality and size of the cloth. Most of the laws concerning weaving do not predate the twelfth century because it was not until then that weaving became an important enough industry to regulate.

Guild regulations, while they hampered innovation and tended to retard guild technology compared to developments outside the guilds, probably had little effect on the development of looms. Only occasionally do we hear of a regulation or dispute that involves a change in the looms. One dispute from 1406 concerns a decree of Henry IV, who ordered that the width of cloth be increased from five-quarters of a yard to six-quarters. This meant that all

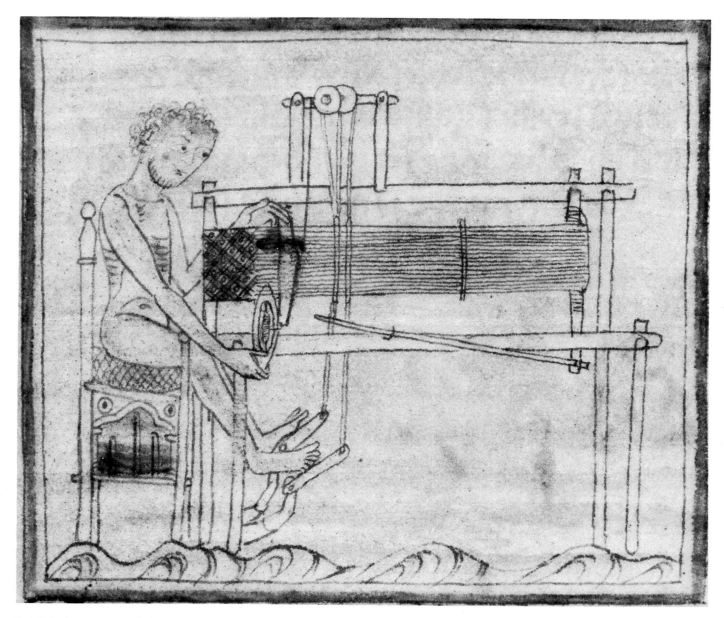

8-4: Naked weaver at treadle loom, c. 1200. Courtesy of the Master and Fellows of Trinity College, Cambridge.

the weavers would have to get wider looms. They protested that they couldn't afford it, and the decree had to be revoked.

EVOLUTION OF LOOM TECHNOLOGY

A second reference (after Rashi) to the change from the warp-weighted to the horizontal loom occurs in an interesting metaphor from the twelfth century by one Alexander Neckam: "Like a knight leaning on two stirrups, the weaver keeps prodding his frugal horse. The pedals of his loom, symbolizing the shifts of his fortune, are pleased in their alternation, so that while one of them rises, the other goes down without the slightest envy." Besides giving us a curious view of medieval horsemanship the metaphor de-

scribes the counterbalance action of the simple two-harness treadle loom. The earliest clear illustration of this new tool appears in a thirteenth-century English manuscript (fig. 8-4).

As crude as the illustration is, it demonstrates that the medieval horizontal loom is essentially the same as the loom used by handweavers today (see fig. 8-32). The warp is stretched between two revolving beams. The warp beam is locked in place by a stick that passes through the end of the beam and is braced against the side rail. This lever, more typical of linen than of wool weaving, provides a certain elasticity in the warp tension that absorbs the strain of changing sheds. The treadles operate a pair of harnesses that are counterbalanced over a pulley above. The superstructure that supports the harnesses also appears to support a reed-beater suspended from a cord. Since the

Middle Ages this loom has been refined to enable it to accommodate additional harnesses, to modify the shedding mechanism, or to make it sturdier, but in essence it has remained unchanged. It characteristically consists of one self-contained, boxlike frame on which all the operations of the actual weaving—treadling, shedding, beating in, advancing the warp—occur.

Not all medieval looms took the form of the loom shown in fig. 8-4. On the late-fourteenth-century loom shown in fig. 8-5, for example, the pulleys for four-harness weaving are suspended from a bar attached to the ceiling. The four posts that support the warp and cloth beams and the pivot bar for the treadles are not part of an integrated frame but are either attached to or embedded in the floor. (The exaggerated balls of yarn in the lower-left corner were common attributes of weaving scenes, perhaps indicating that there was plenty of work at hand.)

The prototype for this loom might be seen in a fourteenth-century illustration of a Byzantine loom from a Greek manuscript of the Book of Job (fig. 8-6). The rudimentary frame consists of just the supports for the harness pulleys. Though it is unclear from the illustration, it appears that the reed was not suspended but rather rested on the warp when not in use. The warp beam is held in the crotches of two posts and secured by a lever much like the one shown in fig. 8-4. The cloth beam, for all that is shown of it, could very well be held by a backstrap arrangement. The weaver here is a woman, suggesting that the cloth woven on this loom was for home use.

Compared to the spectacular development of Flemish and English woolens and Flemish tapestries, the advance of loom technology during the Middle Ages was modest indeed. One historian noted that "weaving was commonly done on looms with two or three treadles and *occasionally with four cloth beams* [my italics]." The next major advance in loom development after the introduction of the treadle loom would not occur until the Industrial Revolution in the eighteenth century.

One notable thirteenth-century development, however, probably of Flemish origin, was the two-man loom (fig. 8-7). The four-harness model illustrated here from the Ypres *Book of Trades* shows a ratchet-and-pawl mechanism for locking the warp beam in place. The heavy, closed-beam construction had become common by the fifteenth century for the support of the suspended reed-batten. It is generally assumed that the child in front is spinning yarn, but, as medieval spinners usually spun standing up, it is more likely that he is winding spools of weft for the boat shuttles.

The horizontal loom moved more slowly into Scandinavia. The evidence of its northern progress is scanty and includes a harness pulley found at Sigtuna, Sweden that dates no later than the thirteenth century. In the fourteenth century new words relating to the horizontal loom began to appear in Danish and Swedish—such as *solv* ("heddle") and *fyrskaft* ("four-shaft" or "four-harness")—which indicate that a transition was occurring. Marta Hoffmann believes that the Norwegian word *ferskeptr*, an adjective describing cloth woven on four harnesses, entered the language early in the fourteenth century. It is doubtful that this new word referred to four-shed weaving on the warp-weighted loom, because that was done with three heddle rods and one natural shed and was known as *priskept*. Hoffmann does not conclude that the appearance of the word *ferskeptr* in the language necessarily implies the appearance of the horizontal loom. Too much uncertainty exists regarding the kind of cloth that the term actually described. It is possible that a more complicated fabric was woven on the warp-weighted loom using four heddle rods. Janice S. Stewart states in *The Folk Arts of Norway* that the horizontal loom was not in general use in Norway until about 1750, but this date seems somewhat late. The earliest dated example of such a loom is from Setesdal and is dated 1668 (fig. 8-8). It is a four-harness loom with an overhead, suspended beater and is clearly a well-developed tool.

Iceland did not begin to import the horizontal loom until the eighteenth century, possibly because Icelandic houses were too small to contain it. Even so the quality of the cloth that Icelandic weavers could produce on the prehistoric warp-weighted loom was fine enough to compete on the European market until the eighteenth century.

The horizontal loom did not arrive alone in Europe; it seems to have been accompanied by a whole complex of related tools—the warping board and mill, the spool rack, the sectional warp beam, the bow, the spinning wheel, and the rotary wheel and cage spools. A cultural loan tended to include the entire constellation of work processes: raw materials, tools, and the methods of working those materials. Many of the new tools were never adopted by domestic weavers; others were adopted only much later. These were professional implements for which little need existed in home weaving. Even among the professionals some of the new tools were regarded with suspicion. For example, wheel-spun yarn was prohibited in Abbeville in 1288. In 1290 a Drapers' Guild regulation at Speyer (Spires) prohibited the use of wheel-spun yarn for the warp but allowed it for the weft. The medieval wheel was more like a mounted spindle than a spinning wheel. It had no flyer until the late fourteenth century and was turned by a hand crank until the development of the crank and connecting rod in the early sixteenth century. Guild members felt that a finer, stronger thread could be spun by the drop spindle, and this simple tool remained the preferred method until the fifteenth century.

The warping board (fig. 8-9) probably derived from the method of warping around pegs pounded into a wall, as was practiced in ancient Egypt (see fig. 3-3). Round

8-5: Medieval treadle loom, painting in Mendel House book from Nurnberg, 1389. The harnesses are suspended from the ceiling. Courtesy of Ciba-Geigy Ltd.

8-6: Byzantine treadle loom from Book of Job miniature, 1368. Courtesy of *Acta Historica*, Budapest.

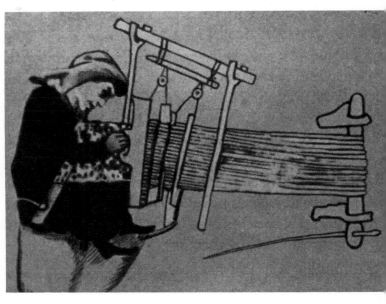

8-7: Two-man loom from the Ypres *Book of Trades*, c. 1310. From DeVigne, *Recherches historique sur les costumes civils et militaires des gildes et corporations de métiers*, 1847.

8-8: Treadle loom from Setesdal, Norway, 1668. Norsk Folkemuseum, Bygdøy-Oslo.

8-9: Spool rack and warping board from the Ypres *Book of Trades*, c. 1310. From DeVigne, op. cit., 1847.

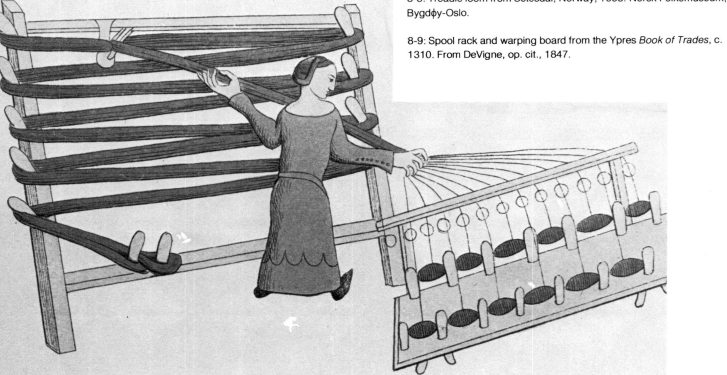

8-10: Iron wool combs, 19th c. Width c. 8½", length of teeth c. 5½". Courtesy of National Museum of Iceland, Reykjavík. Photograph by Gísli Gestsson.

warping frames were also used, and the double pegs at the end for keeping the cross have not changed to this day. The warp was often sized with the liquid from boiled rabbit skins or an adhesive made from the waste of corn mills. Though sectional warping, the spool rack, warping a number of threads at one time for a long length, and making warps into chains all seem related to the new loom, the new tools did not necessarily lead to a change in the basic principles of weaving. Counting threads and weighing the warp were ancient methods that changed, if at all, only in the methods of counting that the new tools required.

The wool comb (fig. 8-10), used in pairs for combing out wool to align the fibers in preparation for spinning, is an ancient tool that dates to the Dark Ages in northern Europe and to classical times in Greece and Rome. The combs generally had one to two rows of iron teeth about four inches long and during medieval times made an effective instrument of torture for raking flesh as well as wool. With the invention of wool cards in the fourteenth century the combs became specialized for long-stapled wool. The early cards set with teasles would not have worked for combing wool, but the stiff yet slightly supple points of the teasle made an ideal instrument for raising the nap on finished cloth. The later cards with bent iron teeth set in leather may also have been invented originally for the purpose of raising the nap on cloth, but the guilds did not permit that use. It was discovered that carding separated the fibers into a spongy mass that made spinning easier, and the "improved" cards were then used to comb short-stapled wool just as they are today.

In the later Middle Ages mercantile control gradually dominated clothmaking. By keeping the artisans dependent on them for raw materials and markets the merchants made it virtually impossible for the weavers to get out of debt and rise into affluence. This trend was epitomized in fourteenth-century Florence, where bankers and merchants had developed devastatingly efficient business practices that eventually reduced the textile worker to a mere cog in the machine.

At the same time wool famines, strikes, repression, and political considerations began to erode Flemish dominance in clothmaking, and by the late thirteenth and early fourteenth centuries, Flemish weavers began to emigrate to Italy and England in search of work. England saw the profits that could be made from textiles and, at the instigation of Edward III, began to develop its own weaving industry.

The Flemish merchants couldn't compete with Florentine capital and business practices. In 1338 there were over two hundred textile workshops in Florence, and over 30,000 workers made their living in the cloth trade. The clothworkers, dependent on the merchants for selling their goods, became chained to these middlemen, a mere two percent of the population, by financial considerations.

During the Renaissance the centers of weaving continued to shift with the political and economic tides. In 1530 the Prince of Orange and Pope Clement VIII overran Florence, and textile production declined. Looms during the same period changed very little (figs. 8-11 and 8-12). Ad-

8-11: Treadle loom from Rodericus Zamorensis, *Spiegel des menschlichen Lebens*, Augsburg 1479. Note how warp is diverted upward, probably to save space in a small house. After Mummenhoff, *Der Handwerker*, 1901.

8-12: Treadle-loom weaver with woman bringing yarn. From *Piazza Universale*, 1641. Courtesy of Ciba-Geigy Ltd.

justments in construction made it possible to hang up to twenty-four harnesses in the loom for complicated patterns, but the major changes took place in fabric design and in the uses of fiber, not in the looms themselves. Two illustrations of Penelope at the loom, both showing the artists' ignorance of the warp-weighted loom of the ancient Greeks, suggest the contemporary two-harness loom with which the artists probably *were* familiar (figs. 8-13 and 8-14). The loom illustrated by Pinturicchio is finely crafted with a boxlike frame, perhaps exaggerated for artistic purposes. (Note how the top bars of the loom guide the eye and frame Ulysses's ship in the background.) Holbein illustrates a more simply constructed loom without the box-like frame, but the two function identically.

Not until the publication of Diderot's *Encyclopédie; ou, Dictionnaire Raisonné des Sciences, des Arts, et des Métiers* (1751–72) did anyone take a serious interest in accurate representations of weaving implements. (An earlier effort by the French Royal Academy of Science, founded in 1666 during the reign of Louis XIV, was begun under the influence of Colbert, the Sun King's minister of state, in 1675. He commissioned the Academy to investigate and describe French industry, but this project languished for

eighty-six years, with plates commissioned but never published. A year before Diderot published his own plates in 1762—many of them allegedly plagiarized from the Academy's collection—the Academy rushed its first volume into print.) Of the eleven volumes of plates published between 1762 and 1772 that accompanied *L'Encyclopédie*, those illustrating the textile arts are among the least satisfactory in detail. Nonetheless they represent the first concerted effort at scientific, technical illustration of the tools of linen, silk, gauze, and velvet weavers, and the results are well worth reproducing here (figs. 8-15, 8-16, 8-17, and 8-18).

The two looms shown in fig. 8-15, the least detailed, depict a variation of the counterbalanced box-frame loom. The treadles have been recessed in the floor, and the cloth beam moved down to the weaver's knees, giving him more room for weaving on top. It is clear that the warp beam is held by friction brakes (ropes attached to the frame at one end and hanging weights at the other encircle the beam), but no device is shown for locking the cloth beam in place. Other views of this loom (fig. 8-16 and more detailed engravings not reproduced here) indicate that the cloth beam is suspended from the side rails by ropes and kept in place by the sheer weight of the beam hanging on the ropes.

The gauze loom shown in fig. 8-17 is an example of a jack-type loom. The jacks sit on top of the loom, pivoting on a rod supported by wooden stanchions. Two sets of lams below appear to be tied up for countermarch action, in which certain harnesses are raised and others are lowered by depressing one or more treadles. The cloth beam is back in the weaver's lap and, as other plates illustrate, is held in place by a ratchet-and-pawl mechanism. Compare this to the silk loom (fig. 8-18), also a jack-type loom tied up for countermarch operation. The silk loom is lighter and more compact. One incongruous detail shows the treadle pivots anchored with a stone instead of being a part of the frame.

Diderot wrote in his article on "Art" for *L'Encyclopédie:* "Let us at last give the artisans their due . . . . It is for the liberal arts to lift the mechanical arts from the contempt in which prejudice has for so long held them, and it is for the patronage of kings to draw them from the poverty in which they still languish." His call for a new order for craftsmen was not without irony. *L'Encyclopédie* was published on the eve of the Industrial Revolution, a chain of inventions and events that would transform the weaver's brief period of glory during the latter part of the eighteenth century into sweatshop labor and economic dependency a few years later.

THE THRUST TOWARD MECHANIZATION
The thrust toward mechanization in weaving did not begin with the Industrial Revolution—roughly the mid-eighteenth

8-13: *The Suitors Surprising Penelope*, fresco by Bernardino Pinturicchio (1455–1513). The artist has seated his subject at a contemporary Italian loom. Reproduced by courtesy of the Trustees, The National Gallery, London.

to mid-nineteenth centuries in England—though it was that period that saw its fullest flowering. Leonardo da Vinci (1452–1519) had experimented with mechanical looms, but nothing came of his efforts. The first successful inventor of an automatic loom is thought to be Anton Möller of Danzig, who had the misfortune to invent a ribbon loom in 1586 that an unskilled person could operate merely by pressing a lever. The measure of his success is that he was ordered strangled (or perhaps drowned in the Vistula) by the City Council for his efforts. Perhaps it was fortunate for us all that da Vinci failed.

According to the ordinances against it in Holland (in 1623, 1639, and 1648) an automatic ribbon loom was used in Leyden perhaps as early as 1621. Some kind of automatic loom, known as a bar loom or Dutch loom-engine, that could weave four to six ribbons at once was reported in London in 1616. By 1621 it apparently had been developed to weave twenty-four ribbons simultaneously. Many cities issued ordinances against the use of such looms in the early seventeenth century, and this loom caused riots in London in 1675 (fig. 8-19).

But the real impetus toward power-loom weaving stemmed from the development of spinning machinery. In the days of handspinning and handweaving, three to five spinners were needed to supply one weaver with sufficient yarn. But, in 1733 another unfortunate inventor by the

147

name of John Kay, an Englishman, invented a device called a *flying shuttle* that could quadruple a weaver's output. Kay barely escaped with his life when his house was stormed by angry weavers in 1753, and eventually he died in poverty in France. Kay's first contribution to weaving lay in substituting metal for cane in reeds. He also invented a spinning machine that was destroyed by spinners who feared for their jobs.

His flying shuttle (fig. 8-20), which is still used today, consisted of a cord with a handle in the middle and metal catching and throwing plates at each end. By jerking the handle the weaver could propel the shuttle (now metal-tipped and set on wheels) along the shuttle race, a widened track along the front of the reed, to the catch box at either end. Not only did the flying shuttle increase the speed of weaving, but also for the first time it allowed a single weaver to weave cloth that exceeded the breadth of his reach. While the flying shuttle was not generally accepted until the 1750s, its effect was to widen even further the disparity between the spinner's and the weaver's production.

Kay's invention led in turn to the series of spinning inventions with which the Industrial Revolution is more commonly associated. Though not the first to attempt it,

8-14: *Penelope at the Loom*, drawing by Hans Holbein the Younger. Marginal drawing in Erasmus of Rotterdam, *Praise of Folly* (Erasmi Roterdami Stultitiae Laus), Basel, 1515. Kunstmuseum Basel, Kupferstichkabinett.

8-15: Two looms in weaving shop. After Diderot, *L'Encyclopédie, Recueil de Planches*, Vol. XI, "Tisserand," Pl. I. The Beinecke Rare Book and Manuscript Library, Yale University.

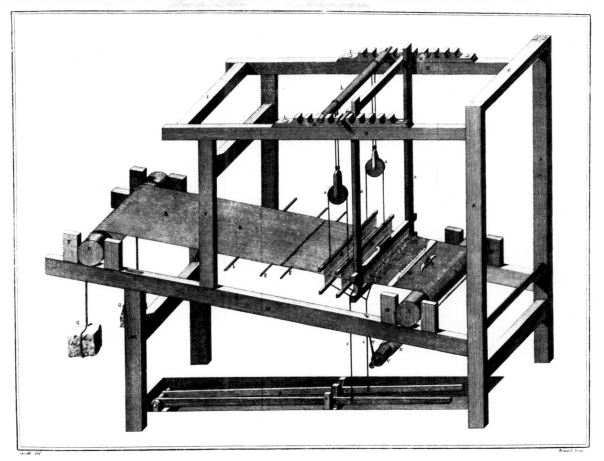

8-16: Treadle loom. After Diderot, *L'Encyclopédie*, *Recueil de Planches*, Vol. XI, "Tisserand," Pl. II. The Beinecke Rare Book and Manuscript Library, Yale University.

8-17: Jack-type loom. After Diderot, *L'Encyclopédie, Recueil de Planches,* Vol. XI, "Métier à Marli," Pl. I. The Beinecke Rare Book and Manuscript Library, Yale University.

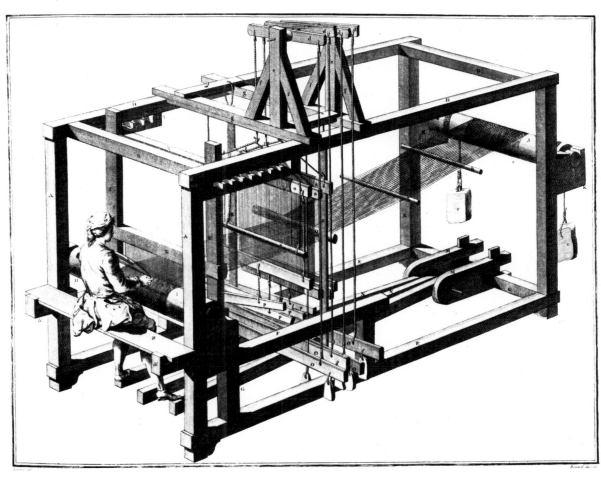

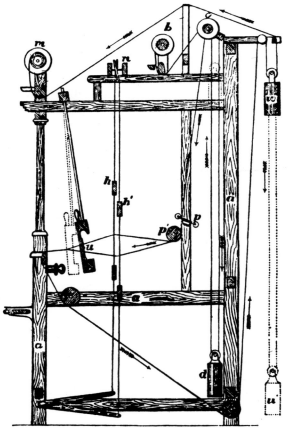

8-18: Velvet loom with spools. After Diderot, *L'Encyclopédie, Recueil de Planches*, Vol. XI, "Soierie," Pl. XCI. The Beinecke Rare Book and Manuscript Library, Yale University.

8-19: Dutch ribbon loom. (b) Warp reel. (c) Pulleys. (d, w) Weights. (p) Beam. (h, h') Heddles. (u) Reed. (m) Cloth roller. The warp follows the direction of the arrows. After Barlow, *The History and Principles of Weaving by Hand and by Power*, 1878.

James Hargreaves is regarded as the first successful inventor of a spinning machine, His *jenny*, c. 1765, named after his daughter, derived from the observation that a spinning wheel knocked on its side continued to turn. Elaborating on that principle, Hargreaves developed a machine that could spin up to eight yarns at once, though they were not strong enough to be used as warp. (He too was driven out of town by an angry mob that destroyed his machines.)

By 1769 Richard Arkwright had improved on the jenny by adding successive pairs of rollers, each set revolving faster than the previous set, that drew out the yarn, twisted it, and wound it on bobbins in one continuous action. Arkwright's machine, powered at first by a horse, was later adapted to water power and still later to steam. His *water frame,* as it was called, produced a cotton thread strong

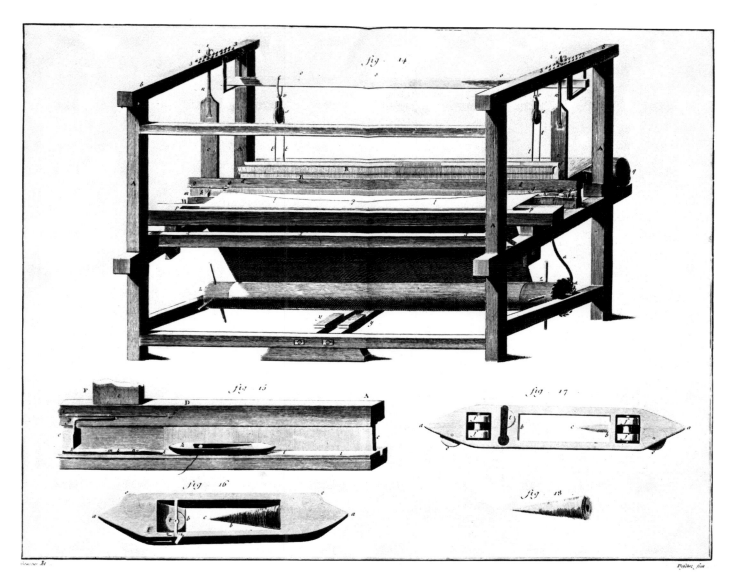

8-20: Loom with flying shuttle. After Diderot, *L'Encyclopédie, Recueil de Planches*, Vol. III, "Draperie," Pl. IV. The Beinecke Rare Book and Manuscript Library, Yale University.

enough for the warp, and, although he too was attacked by the handspinners, he was more fortunate than his predecessors: he was a clever businessman and died a millionaire. Arkwright is credited with establishing the world's first cotton spinning mill in 1769 in Nottingham. His invention also marks the beginning of the factory system in England and along with it the beginning of child labor.

Ten years later Samuel Crompton, who by now should have learned the risks of innovation in the textile industry, produced what was jokingly called a spinning *mule*. He combined the best aspects of Hargreaves' jenny with the best of Arkwright's horse-powered frame and came up with a machine that was the first to produce a thread fine enough and strong enough for calicoes. Wary of having his machines destroyed by spinners, he sold his mule to a group of manufacturers for promises of wealth that

remained just that—promises. Not only was he never paid for his ideas, but his life was threatened, and, as the manufacturers grew rich, Crompton lived in poverty on an annuity provided by his friends.

The pendulum now swung the other way. Between 1770 and 1800 cotton consumption soared twelve times, and handweavers were hard put to keep up with yarn production. Thanks to spinning machines and the fly shuttle, the weaver suddenly became a prosperous and respected craftsman. This was, as David S. Landes put it in *The Unbound Prometheus*, "the golden age of the handweaver, whose unprecedented prosperity was a shock to all, a scandal to some." It was not to survive the third decade of the nineteenth century.

The pressure now mounted for a mechanical weaving machine that could keep pace with the spinners. There had been several unsuccessful attempts earlier, most notably that of de Gennes in 1678 at a loom powered by water that

151

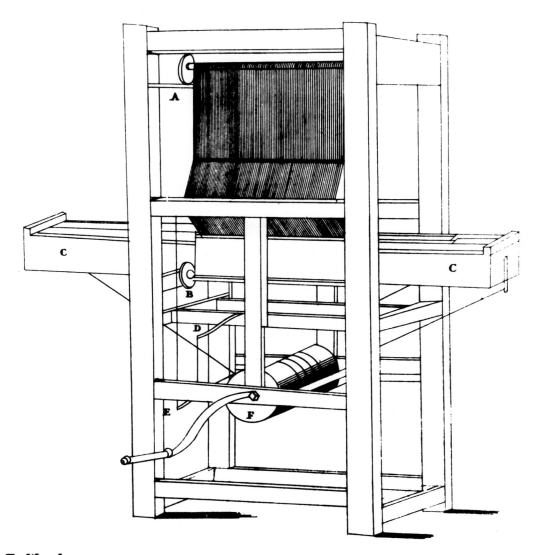

A *The Warp beam*

B *The Cloth beam*

C C *The boxes containing the springs that throw the shuttles.*

D *A lever, having a corresponding one on the opposite side, for elevating the reed or comb.*

E *A lever, having a corresponding one on the opposite side, for reversing the threads.*

F *The Cylinder, which gives motion to the levers.*

N.B. The warp is kept to a due degree of tension by the counteraction of either a weight or spring.

The web is made to wind by the like power, tho'   in an inferior degree; and is prevented, as the stroke of the reed or comb brings it down from unwinding by a ratch wheel and click.

The enrolled drawing is colored.

Drawn on Stone by Malby & . . .

8-21: Cartwright's loom, 1785. National Museum of Natural History, Smithsonian Institution.

worked the heddles with cams and another by John Kay in 1745. The solution—and perhaps the most curious addition to this unusual chain of inventions—was arrived at by an Anglican clergyman named Edmund Cartwright, who had never even seen weaving performed and who invented his loom on a bet. So remarkable was his achievement that Cartwright should be allowed to describe it in his own words:

> . . . it struck me that, as in plain weaving, according to the conception I then had of the business, there could be only three movements which were to follow each other in succession, there would be little difficulty in producing and repeating them. Full of these ideas, I immediately employed a carpenter and smithy to carry them into effect. As soon as the machine was finished, I got a weaver to put in the warp, which was of such material as sail cloth was usually made of. To my great delight, a piece of cloth, such as it was, was the production.
>
> As I had never before turned my thoughts to anything mechanical, either in theory or practice, nor had even seen a loom at work, or knew anything of its construction, you will readily suppose that my first loom must have been a most rude piece of machinery. The warp was placed perpendicularly, the reed fell with a force of at least half a hundred weight and the springs were strong enough to have thrown a Congreve rocket. In short, it required the strength of two powerful men to work the machine at a slow rate and only for a short time. Considering in my great simplicity that I had accomplished all that was required, I then secured what I thought most valuable property by a patent, April 4, 1785 [fig. 8-21]. This being done, I then condescended to see how other people wove; and you will guess my astonishment when I compared their easy mode of operation with mine. Availing myself, however, with what I then saw, I made a loom, in its general principles, nearly as they are now made; but it was not till the year 1787 that I completed my invention, when I took out my last weaving patent, August 1 of that year.

Cartwright powered his first loom by an ox and capstan but soon adapted it for steam. (James Watt had patented his steam engine in 1769.) Cartwright's loom initially suffered the same fate as that of previous textile inventions. He had trouble interesting manufacturers in his unusual design, so he set up his own factory at Doncaster. But Cartwright was no businessman, and after nine years and £30,000 he gave it up, selling some of his looms to a Manchester firm. Angry weavers burned the Manchester factory and four hundred of his looms to the ground.

Workers protesting and rioting over the loss of jobs created havoc in early nineteenth-century England, but there was no stopping the surge toward industrialization. Cartwright's loom caught on, and by 1833 one man with a twelve-year-old assistant could operate four looms and produce twenty times the output of a handweaver. The golden age of the handweaver was over in England, and the artisans sank back into the obscurity in which Diderot had found them in the mid-eighteenth century. They had nothing to do with the products of industrialization, and the term "machine-made" became synonymous with bad taste and inferior quality. Home weaving, of course, continued as before on looms that varied little from those used during the Renaissance. Some had a box, or partial box, frame (fig. 8-22); others were built with uprights and castle to support the harnesses and beater (fig. 8-23). It was this home weaver's loom, a rugged, counterbalance-type loom for weaving linens and woolens, with revolving beams and suspended, overhead beater, that was introduced into colonial America in the early seventeenth century.

THE COLONIAL LOOM

Although no looms came over on the *Mayflower*, the ship did carry some weavers, among them William Bradford, a fustian weaver from Austerfield, England and governor of the Plymouth Colony for most of his American life. During the early colonial period Britain supplied most of the colonists' textiles while encouraging them to plant flax and hemp, and even to cultivate silkworms—both for their own textile needs and eventually for export. In 1623, for example, the Virginia legislature ordered each settler to plant mulberry trees for silk cultivation, at least one for each ten acres of land. But the settlers, struggling to hack an existence out of the wilderness, had many demands on their time and found tobacco an easier and more profitable undertaking.

Of necessity almost every colonial home became a miniature textile manufactory. While not every home had a loom, virtually every home did have a spinning wheel and a patch of flax growing in back. Sheep had arrived with the first settlers at Jamestown in 1609, but they produced an inferior wool. Not until the early nineteenth century were Merino sheep imported to improve the breed. Cotton cloth was woven as early as 1642 from West Indian cotton, but since it took a whole day to pick the seeds from a single pound of cotton, raising it on a large scale would have to wait for another hundred years or so until slavery made it economical. Early in the seventeenth century the colonists realized that they would have to become self-sufficient for their clothing needs, and colonial legislatures began to offer bounties for growing hemp and flax, raising sheep, and killing wolves. Perhaps the bounties were not a sufficient incentive, for some legislatures passed laws mandating that each household produce a certain amount of spun yarn each year.

8-22: *The Loom*, drawing by Vincent Van Gogh, 1884. Rijksmuseum, Kröller Müller, Otterlo.

By the end of the century Great Britain, which earlier had encouraged colonial textile manufacturing as a possible source of raw materials for the mother country, began to recognize that that very manufacturing might pose a threat to her own textile trade. The threat of colonial woolens, for example, inspired this law in 1699:

That from and after the first day of December, in the year of our Lord one thousand six hundred ninety-nine, no wool, woolfells, shortlings, mortlings, wool-flocks, worsted, bay, or woollen yarn, cloth serge, bays, says, frizes, druggets, cloth-serges, shalloons, or any other drapery stuffs or woollen manufactures whatsoever, made or mixed with wool or woolflocks, being a product or manufacture of any of the English plantations in America, shall be loaden on board any ship or vessel, in any place or parts within any of the said English plantations, upon any pretense whatsoever.

Needless to say, British restraints on trade only stimulated colonial textile manufacturing. Wearing home-spun quickly became a patriotic act—so much so that in 1767 Governor Moore of New Jersey was able to report: "The custom of making these coarse clothes (woolen and linsey-woollen) in private families prevails throughout the

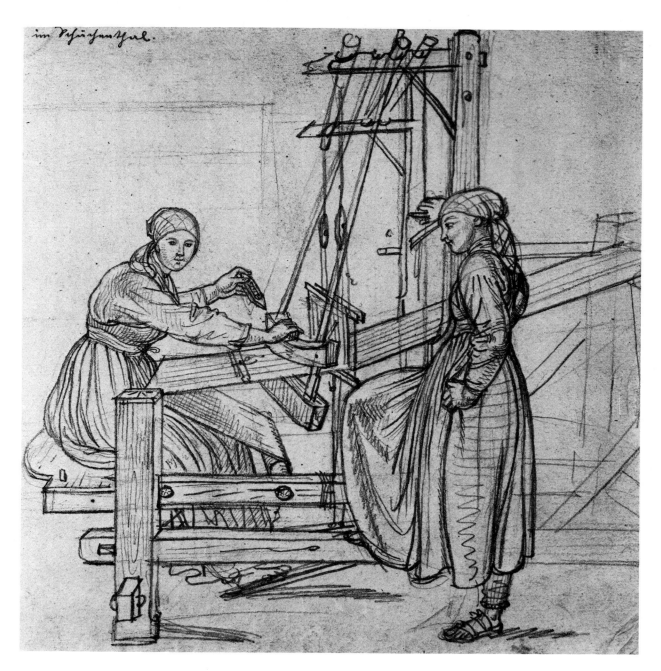

8-23: Home weaving in the canton of Uri, Switzerland. Drawing from the first half of the 19th century by Ludvig Vogel (1788–1879). Swiss National Museum, Zurich.

whole province, and in almost every house a sufficient quantity is manufactured for the use of the family. . . . Every home swarms with children, who are set to work as soon as they are able to spin and card. . . .'' In that same year, the graduating class of Rhode Island College (now Brown University) wore only homespun clothes. Harvard followed suit a year later.

The kinds of textiles that the colonists wove included plain linen, woolen, and cotton cloth; mixtures of wool and flax, cotton and flax, cotton and wool (linsey-woolsey), fustian, and jeans; tow cloth from the coarsest fiber of flax; ducking from hemp, chiefly for sailcloth; coverlets and counterpanes; and carpets. On what type of loom was this impressive array of fabric woven?

The colonial loom had a mixed ancestry, stemming as it did from the Dutch in New Amsterdam, the Mennonites in Pennsylvania, the Scotch in the South, the Irish in New Hampshire, and the English Puritans in New England. No single loom could be singled out as *the* colonial loom, for they varied in detail from place to place. The typical colonial loom, however, was a four-post loom. The four

8-24: Colonial loom with overhead beater, height 72½″, covering floor space 60½″ x 64½″. Collection of the Newark Museum.

square timbers stood about seven feet high and as far apart as the posts of a bed (fig. 8-24). The warp beam was made from a close-grained, well-seasoned wood and was about six inches in diameter, with an iron axle driven into it before it was turned on a lathe. The cloth beam was about ten inches in diameter.

Another example of a four-post loom (fig. 8-25) from a nineteenth-century book of trades was equipped with lams (to raise more than one harness by depressing a single treadle) and braced at the top to ceiling beams. The car-

pentry on the early colonial looms, when each family did everything for itself, must have been crude. Later a loom maker might charge eight to ten dollars for the carpentry, but craftsmanship had to wait until communities grew large enough to permit specialization.

The loom may have been placed on a side porch or in an attached shed, but more often it stood in a corner of the kitchen where the busy housewife could get to it conveniently in her few odd moments of leisure during the day. As Richardson Wright quaintly described it in *Hawkers and Walkers in Early America*, "In those days there was the whiz of the shuttle, the jarring of the lathe, and the clattering of the treadles, while buzz-buzz went the rapid wheel

156

and creak-creak the windle from which ran the yarn the rosey daughter was quilling."

Another popular loom design is illustrated in fig. 8-26, a reconstruction of a nineteenth-century Pennsylvania loom of a type used in the colonies since the seventeenth century. The frame is shorter than that of the four-post loom, and the seat, built into the frame of the four-post loom, is here freestanding. Compare this to the loom in fig. 8-27, also probably from Pennsylvania (1789), in which the design is reversed. Here the front posts support the beams from which the harnesses and reed are suspended, and the seat *is* built in. The cloth does not go around the breast beam as in fig. 8-26: it goes *through* it. Finally, note the friction brakes, similar to those in Diderot's illustrations, for holding the warp beam in place. The sturdy beams from which this loom is constructed ably illustrate an old Appalachian saying, "The heavier the loom, the lighter for the weaver."

The reed may have been made of metal or thin strips of reed, with perhaps fifty to sixty dents per inch for fine cloth. The yarn was wound onto corncob spools or quills made from thistle stems with the pith removed. These spools were placed on a spool rack, or skarne, and wound onto a warping board in bouts of forty threads at a time. The warping board may have been a plank about six feet by one foot or just as often merely pegs knocked into the side of a barn. A raddle (also called a rake, ravel, or wrathe) kept the bouts from tangling during the "thumbing in" or threading of the heddles.

By the time of the Revolution, three kinds of weavers populated the colonies—home weavers, weavers who had established shops in the towns, and, perhaps most interesting of all, itinerant weavers. This latter group traveled from homestead to homestead, some bringing their looms with them, others weaving on the household loom for which the farmer's wife had insufficient time, experience,

8-25: Colonial loom from Hazen, *The Panorama of Professions and Trades or Every Man's Book*, 1836.

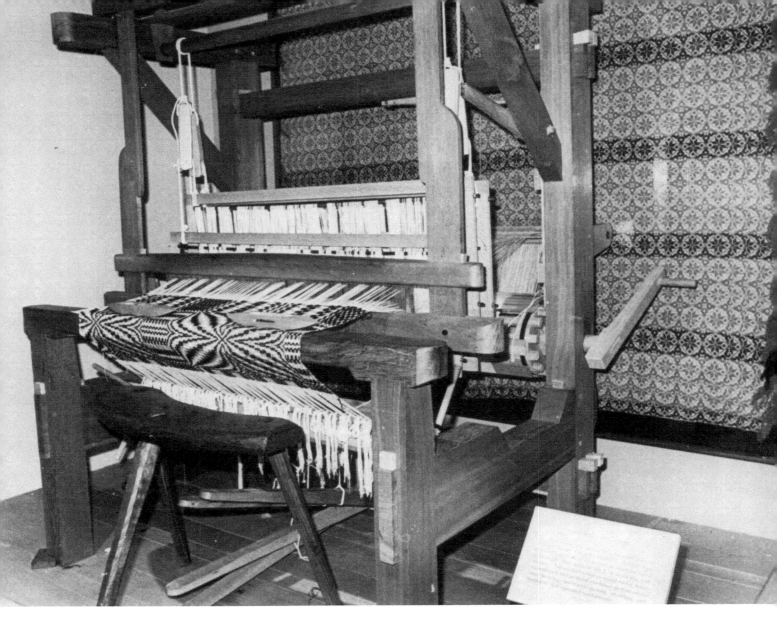

8-26: Twentieth-century reconstruction of a 19th-century loom from Pennsylvania. National Museum of Natural History, Smithsonian Institution.

or interest. With roads and communications in a rather primitive state, the itinerant weaver was always a welcome source of news and gossip. When he finally settled down, his shop became, like the country store of later times, the town center for scandal, rumors, and news.

The Revolution itself inspired some extraordinary accomplishments at the loom. One account tells of a New Hampshire woman who sent her brother off to war in a woolen suit that she had sheared, washed, carded, spun, and woven in twenty-four hours. When the Provincial Congress cried for 13,000 warm coats for Continental soldiers, hundreds of women sat down and spun and wove them. Each soldier who enlisted for at least eight months received one of these ''bounty coats'' and prized it highly, for they seemed to last forever.

After the war the new country went through a short-lived craze for foreign finery, which had become cheap and accessible for a time, but an embargo by Jefferson, westward expansion, and perhaps overspending on European goods soon sent the settlers back to their looms. But the age of homespun was drawing to a close. In 1775 in Philadelphia the first spinning machine made its appearance in America, with a capability of spinning simultaneously twenty-four threads of cotton or wool. In 1788 in Providence a Scotchman by the name of Joseph Alexander introduced the flying shuttle to the American hand-loom. In 1790 the first carding mill was established in Pittsfield, Massachusetts, and in the same year young Samuel Slater, having memorized the construction of Arkwright's spinning machinery in England, opened his spinning mill in Pawtucket, Rhode Island. One invention chased the next. In 1793 Eli Whitney invented the cotton gin, and in the first seven years of its use cotton production soared a hundredfold. Suddenly slavery became practical.

8-27: Early American loom, probably from Pennsylvania, 1789.
Merrimack Valley Textile Museum.

8-28: *Little Rocking Loom.* Courtesy of Bigelow-Sanford, Inc.

The first weaving "factory" had been established in 1638 by Ezekial Rogers, who brought twenty Yorkshire families over from England to weave woolens and fustians in Rowley, Massachusetts, but the factory as we know it did not make an impact on American textile production until the growth of industrialism at the turn of the nineteenth century. The power loom was introduced by William Gilmore at Waltham, Massachusetts in 1815, and in 1825 William H. Horstmann of Philadelphia imported the first jacquard loom. By 1830 much of a family's spinning and weaving needs was machine-made.

Between 1815 and 1830 the price of ordinary brown shirting fell from forty-two to seven and a half cents a yard. Housewives tramping out four yards of cloth a day on their handlooms couldn't compete with mill production, in which one man tending three or four looms could produce ninety to one hundred sixty yards a day. Industrialization killed the incentive for home weaving, but the expanding frontier kept home weaving alive until about 1860 when as an economic factor it finally disappeared.

The South experienced a revival of household production during the Civil War, but by then the era of handwoven coverlets and counterpanes, perhaps the greatest expression of early American weaving, had come to a close. Almost. It continued in the isolated Appalachian regions of the South until the end of the century. It was dying out even there in the 1890s when a small revival began at Berea College in central Kentucky under the presidency of William Goodell Frost. Frost recognized the value of this dying art and swapped education for coverlets, which he marketed in Boston. The revival spread to North Carolina, where Frances L. Goodrich set up Allenstand Cottage Industries in 1895, and with varying amounts of success the revival continued into the 1920s. The last word on Appalachian weaving belongs to Aunt Sal Creech of Pine Mountain, Kentucky, a weaver of coverlets: "Weaving, hit's the purtiest work I ever done. It's settin' and trompin' the treadles and watchin' the blossoms come out and smile at ye in the kiverlet."

Sometime during the nineteenth century loom design underwent one further alteration. The beater, which had swung from an overhead bar (the "rocking tree," as it was known in colonial America) since medieval times, was flipped upside down and pivoted on pins in the lower side bars (fig. 8-28). This simple design change eliminated much of the heavy superstructure needed to support the beater, and the loom became a smaller, more compact tool that could be used in almost any room of the house, including the attic. It was this loom, called "the little rocking loom" in the Southern Highlands of Appalachia, that became the prototype for the contemporary treadle loom used by most handweavers today.

THE CRAFT REVIVAL AND
THE CONTEMPORARY LOOM

THE ARTS AND CRAFTS MOVEMENT

The revival of hadweaving in the United States and Great Britain has its roots deep in the Industrial Revolution. As early as 1829 Thomas Carlyle expressed his suspicions about mechanization: "Our old modes of exertion are all discredited, and thrown aside. On every hand the living artisan is driven from his workshop, to make room for a speedier inanimate one. The shuttle drops from the fingers of the weaver, and falls into iron fingers that ply it faster. . . . Men are grown mechanical in head and heart, as well as in hand. They have lost faith in individual endeavor, and in natural force of any kind."

Carlyle's lament for lost craftsmanship was echoed by John Ruskin, the English essayist, critic, and reformer, who proclaimed, "Industry without art is brutality." Both Ruskin and Carlyle feared the effects on the human soul of denying the worker the satisfaction of seeing his product through from start to finish. (Their concern, 150 years ago, has a depressingly modern ring to it.) For his part Ruskin championed the cause of handweavers in England, encouraging weavers on the Isle of Man to produce wool guaranteed to "last forever." He helped revive cottage industries in Westmorland and Cumberland and influenced young William Morris with the notion that work should be pleasurable.

William Morris (1834–96) is often credited with almost singlehandedly reviving the handcrafts in England. He was trained as an architect and painter and believed along with Ruskin that beauty was essential to survival. When he started to set up his studio in 1857, he found to his horror that he couldn't procure furnishings that suited his aesthetic ideals. In reaction to the shoddy craftsmanship that he found around him and believing that an artist, to produce fine work, needed fine work around him, he set up his own company in 1861 to revive the excellence in craftsmanship that had existed during medieval times.

Morris was not interested in creating art for industry but in creating craftsmen as artists. The ideals of the medieval craft guilds were his guiding principles. His advice to the homemaker was, "Have nothing in your houses which you do not know to be useful or believe to be beautiful." With protean energies the new firm of Morris, Marshall &

Faulkner, Fine Art Workmen in Painting, Carving, Furniture, and the Metals, launched its campaign, and its influence was widely felt throughout all the applied and decorative arts. Unfortunately, the movement that Morris inspired severed the craftsman from industrial development. Yet, he can hardly be faulted for this, for by doing so he probably saved him from extinction.

This movement acquired the name Arts and Crafts when some younger members of the Royal Academy, seeking greater recognition for the applied and decorative arts, founded in 1888 the Arts and Crafts Exhibition Society in London to exhibit handicrafts. The movement spread rapidly, spawning five societies in England between 1880 and 1890. It proliferated in Europe as well, particularly in Scandinavia, where the industries, still craft-based, had not been devastated by the Industrial Revolution as they had elsewhere in Europe. The Scandinavian influence grew as increasing numbers of craftsmen flocked there to study.

In the United States the state of the applied and decorative arts at the end of the nineteenth century was much the same as in Europe. Oscar Wilde, on a lecture tour of the United States in 1882–83, commented on the debased state of the applied arts, implying that Americans were even less sensitive to the situation than the British. "I find what your people need," Wilde said, "is not so much high, imaginative art, but that which hallows the vessels of every-day use . . . the handicraftsman is dependent on your pleasure and opinion. . . . Your people love art, but do not sufficiently honor the handicraftsmen." But the influence of Morris and the Arts and Crafts movement was well received in the United States, probably even more so than in Europe. The Boston Arts and Crafts Society was founded in 1897, and before long similar groups were springing up all over the country. *Handicrafts* magazine published its first issue in 1902, and *The Craftsman*, perhaps the chief exponent of the movement, published continuously from 1901 to 1916.

The severing of the craftsman from industry might have been complete were it not for a countertrend that found an influential spokesman in the architect Walter Gropius and a home in The Bauhaus School of Design that Gropius founded in Weimar, Germany in 1919. Gropius, influenced by Ruskin and Morris and impressed by the superb results of cooperating craft guilds in constructing medieval cathedrals, wanted to create "a new guild of craftsmen" in the medieval tradition. At the Bauhaus he thus discarded the distinction between student and teacher in favor of apprentices, journeymen, and masters. Unlike Morris, who detested industry and the degradation of its products, Gropius sought an alliance with industry. He recognized that the machine had become a fact of life and that, to realize Morris' ideal of bringing beauty to the common people, he and other artists had to infuse industry with art. The Bauhaus struggled against dissension from within and

8-29: Combination jack, counterbalance, and countermarch loom for fabric and tapestry weaving, 40″ weaving width. Manufactured by Thought Products, Inc., Somerset, Pa. Photograph by H. O. Navratil.

political persecution from without, and in 1933 it was finally closed down by Nazi stormtroopers. To the lasting benefit of the arts in the United States a great many of the artists and craftsmen working at the Bauhaus emigrated here to continue their work and teaching.

THE HANDLOOM TODAY

The combined influence of Morris, the Arts and Crafts movement, and the Bauhaus toward making utilitarian objects beautiful as well as functional found expression in loom design as well as in textiles themselves. For the first time in the almost 10,000 years that man has been weaving artists turned their attention to the tools on which their most

exquisite fabrics were woven. Their rationale could have come from Morris himself, who, commenting on the crux of the Arts and Crafts movement, said, "To give people pleasure in the things they must perforce use, that is one great office of decoration; to give people pleasure in the things they must perforce make, that is the other use of it."

As the revival of handweaving gained momentum following World War II, loom design began to fulfill Morris' credo. From Sweden, with its thousand-year-old tradition of home weaving, came looms that were both mechanically efficient and artistically unsurpassed. The contemporary loom, carefully constructed of the finest straight-grained woods and meticulously finished, even without a warp on its beams, has become an object commanding admiration like any other well-designed piece of furniture (figs. 8-29, 8-30, and 8-31). Three different kinds of floor

8-30: A 45″ four-harness jack loom. The sides are made of three-ply maple, each ½″ thick, glued and pressed under 40 tons of pressure. Courtesy of Currier Heritage Looms, Northwood, N. H.

8-31: A 45″ oak jack-type loom. Courtesy of Putney Mountain Looms, Putney, Vermont.

8-32: A 45″ counterbalanced loom (''Fanny''). Made by Leclerc, L'Islet, Quebec, Canada.

8-33: (a) Formation of the shed on a counterbalance loom. (b) Formation of the shed on a jack loom.

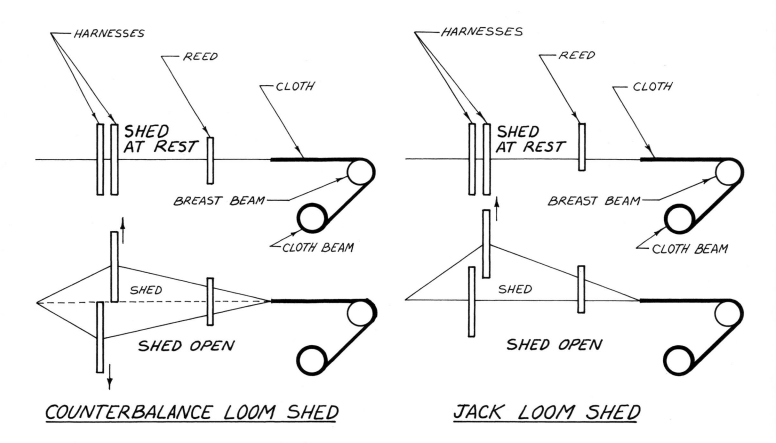

COUNTERBALANCE LOOM SHED

JACK LOOM SHED

looms are commonly used by handweavers today: the counterbalance loom, the jack loom, and the counter-march (or contremarche) loom. Each offers certain advantages and disadvantages to the weaver.

The counterbalance loom (fig. 8-32), the direct descendant of the medieval treadle loom, is the simplest of the three. Like the medieval loom, it operates on the principle of two harnesses counterbalanced against each other over a pulley or roller. The treadles can be connected directly to the bottom of the harnesses or, alternatively, connected indirectly by means of lams. When a treadle is depressed, one harness is pulled down while its mate goes up. This type of loom generally uses only one or two pairs of harnesses, though some looms can accommodate up to four pairs.

Until recently the most attractive advantage of the counterbalance loom over the two other looms has been its relatively low cost. The savings realized by its simplicity in construction, however, have been increasingly eroded by the rising costs of raw materials. While still cheaper than the other two looms, the price tag of the counterbalance loom may no longer be a determinative consideration.

Some authorities have noted that the counterbalance loom is well suited for delicate fibers, because changing the shed places little strain on the warp. When the warp is at rest, it passes through the center of the reed. With each depression of the treadle the warp is raised (or lowered) only half the height of the reed (fig. 8-33). As each shed is changed, all the warps move the same distance either up or down, maintaining a constant, even tension on all the threads. Others have argued that this advantage is more theoretical than real. The counterbalance action also places little strain on the weaver, because the weight of the falling harness helps lift that of the rising harness.

The major disadvantage of the counterbalance loom is its lack of versatility. A weaver would find it difficult, for example, to weave double cloth, which requires lowering three harness while raising one, without the aid of special devices. Since the counterbalance system does not permit three harnesses to be raised or lowered in tandem, the two harnesses containing the warp threads for the layer of cloth *not* being woven with each throw of the shuttle would have to remain at rest through the center of the reed. These warps, unable to be moved out of the way, would frustrate the weaver by reducing the size of the usable shed by half.

Each harness on the jack loom operates independently, making it the most versatile loom of the three. Most jack looms contain between four and twelve harnesses, but twenty are not uncommon. The loom uses a rising shed, and the harnesses can be raised, depending on the construction of the loom, either from below or from above (fig. 8-34). If the jacks push up the harnesses from below, no superstructure is needed at all and the loom can be built low and compact.

On some jack looms the harnesses are suspended from above, but in most cases they sit on a block attached to the inside of the uprights and slide up and down in tracks (fig. 8-31). The clacking of the harnesses back into place when a treadle is released makes the jack loom noisier than the others, but not all weavers consider this a disadvantage. A set of lams permits any combination of harnesses to be tied to any treadle—or combination of treadles—for a great variety of weaves.

The warp on a jack loom rests along the shuttle race at the bottom of the reed. When a treadle is depressed, the shed opens the entire height of the reed (fig. 8-33), with the warps stretching twice the distance that they do on the counterbalance loom. A greater strain is placed on delicate fibers, but for most weavers the versatility of the jack loom more than compensates for this disadvantage. The harness frames can be readily removed for adding additional heddles, and the tie-ups for complicated weaving patterns can easily be made. It is the most popular type of floor loom in the United States.

The countermarch loom is often described as a combination of the counterbalance and jack looms, but the description is misleading. It has no counterbalance action over a pulley or roller, but it does employ both rising and sinking harnesses. The action is achieved with two sets of lams (fig. 8-35). The longer set (b) operates the jacks on top of the loom (c), which raise certain harnesses when a treadle is depressed. The shorter set of lams (a) are tied both to the longer lams and to the bottom of the harnesses in such a way that, when a treadle is depressed, the harnesses that are not raised are pulled down.

Because of the two sets of lams, the countermarch loom requires more space between the treadles and the harnesses than the jack or counterbalance loom. The typical countermarch loom is large, sometimes with the bench built into the frame, and often uses an overhead beater (fig. 8-36). While considerably more complicated to tie up than the jack loom, the countermarch loom offers the same versatility of harnesses that operate independently for weaving unbalanced patterns. The shed opens in the same way as on the counterbalance loom, both up and down from the horizontal plane of the warp, giving the countermarch loom some of the advantages of both the others.

Several types of jacks can be found on countermarch looms. In fig. 8-35 it is clear that, when the jacks at c lift a harness, the bar, pivoting at d, does not lift the harness straight up but tugs it to the left. Pulling the harness out of alignment creates extra strain on the warp. To avoid this problem, two other lifting arrangements are commonly used. In fig. 8-37 the harness is lifted by a cord or cords passing down through the warp to the lams below. This method lifts the harness evenly, but the cord passing through the warp can cause some wear on adjacent threads. The arrangement in fig. 8-38 circumvents both

8-34: *Above:* A 48″ jack loom with overhead jacks. Made by L. W. Macomber Co., Me. Photograph by Barbara Wrubel. *Below:* A twelve-harness jack loom ("Nilart") with jacks that raise harnesses from below. Made by Leclerc, L'Islet, Quebec, Canada.

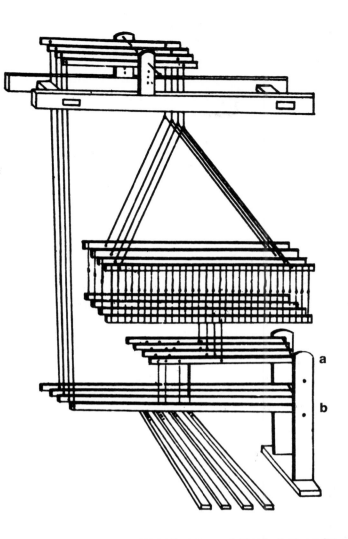

a

b

8-35: Diagram showing how a countermarch loom is tied up. Courtesy of Ulla Cyrus-Zetterström.

8-36: Contemporary countermarch loom. Made by Glimåkra, Sweden. Photograph courtesy of Looms 'n Yarns, Berea, Ohio.

problems by routing the lifting cord outside the warp. When the treadle is depressed, the vertical bar at the top of the loom is pulled to the right, and the ends of the harnesses, tied to the lever at points equidistant from the pivot, are raised evenly.

The variety of looms commercially available to the weaver today is enormous—frame looms, rigid-heddle looms, inkle looms, tapestry and rug looms, table looms, and floor looms. They come in all shapes and sizes for weaving textiles in widths that vary from an inch to five feet or more. They are manufactured by large companies in factories by machine, by small groups of craftsmen in family-size operations, and by individuals with a few power tools and a lot of energy. They come in a range of the finest hardwoods, from cherry, hardrock maple, white birch, and oak, to the softer pines and firs. There are even places that will sell the aspiring craftsman plans to construct his own loom.

If the history of the handloom has been neglected, it is because the loom has never been an end in itself but a means to an end. The loom evolved first as a function of the fibers available for weaving and later as a function of changing fashion. Its evolution is no less significant than that of the printing press, for its development in the various cultures has broadened the expression of their art and ideas. It must be remembered, however, that the loom is merely a machine programmed to do what the artist with the shuttle demands. As Charles Amsden wrote in *Navaho Weaving*: "Unthinkingly, one supposes that our modern machines produce endless variety in textiles; not so, they produce only monotony. The variety is in the machines, not the product, for each machine does just one thing over and over, until its rigging is changed to a new pattern. Fortunately, though, we have many machines at work." Even more fortunately we have many artists at work. Behind the loom we have the imagination of the artist and the skilled hand of the craftsman. One need only recall the "woven wind" of the Hindu or the tapestries of pre-Columbian Peru to recognize that the true limitations of the loom lie not in the machine but in the hand and the eye of the artist.

8-37: Diagram of countermarch tie-up. Courtesy of Ulla Cyrus-Zetterström.

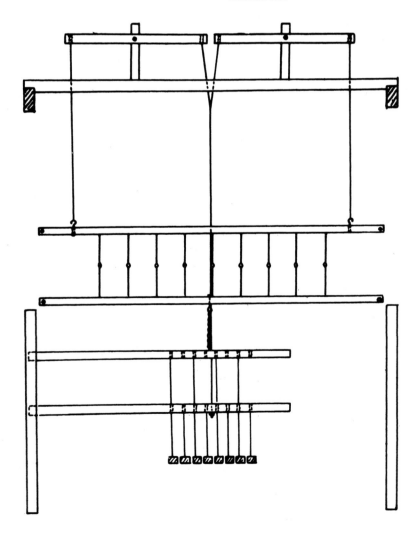

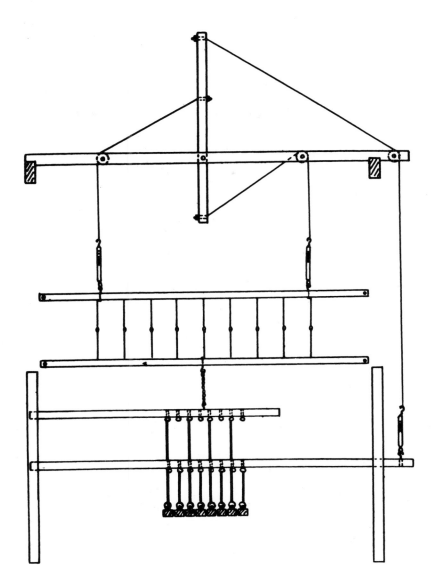

8-38: Diagram of countermarch tie-up. Courtesy of Ulla Cyrus-Zetterström.

Bibliography

Ackerman, Phyllis. "Persian Weaving Techniques. A History." *A Survey of Persian Art*, Vol. 3, edited by A. U. Pope. London: Oxford University Press, 1939.

Adovasio, James M. and Lynch, Thomas F. "Preceramic Textiles and Cordage from Guitarrero Cave, Peru." *American Antiquity* 38 (1973), 84–90.

Aitken, Barbara. "A Note on Pueblo Belt-Weaving." *Man* 49 (1949), Art. 46.

Albers, Anni. *On Weaving*. Middletown, Conn.: Wesleyan University Press, 1965.

American Federation of Arts. *Threads of History*. American Federation of Arts, 1965.

Amsden, Charles Avery. *Navaho Weaving: Its Technic and History*. 1934. Reprint, Glorieta, N. M.: The Rio Grande Press, 1974.

_____ . *Prehistoric Southwesterners from Basketmaker to Pueblo*. Los Angeles: Southwest Museum, 1949.

Andersen, Poul. "Forms and Names of Heddles." *Folk-Liv* 14–15 (1950–51), 60–86.

Atwater, Mary Meigs. *The Shuttle-Craft Book of American Hand-Weaving*. Rev. ed. New York: Macmillan, 1956.

Baines, Edward. *History of the Cotton Manufactures in Great Britain*. London: H. Fisher, R. Fisher, & P. Jackson, 1835.

Baldwin, Gordon C. "Prehistoric Textiles in the Southwest." *The Kiva*, Vol. 4, No. 4 (1939).

_____ . "Survey of Southwestern Prehistory." *The Kiva*, Vol. 6, No. 8 (1941).

Barlow, Alfred. *The History and Principles of Weaving by Hand and by Power*. Philadelphia: Henry Carey Baird & Co., 1878.

Bellinger, Louisa. "Craft Habits, Part 1: Loom Types Suggested by Weaving Details." *Textile Museum, Workshop Notes*, No. 19 (1959).

_____ . "Textile Analysis: Early Techniques in Egypt and the Near East [3 parts]." *Textile Museum, Workshop Notes*, Nos. 2, 3, 6 (1950–52).

_____ . "Textile Analysis: Developing Techniques in Egypt and the Near East [Parts 5 and 6]." *Textile Museum, Workshop Notes*, Nos. 15, 16 (1957).

_____ . "Textile Analysis: Pile Techniques in Egypt and the Near East [Part 4]." *Textile Museum, Workshop Notes*, No. 12 (1955).

_____ . "Textiles from Gordion." *Needle and Bobbin Club, Bulletin* 46 (1962), 4–33.

Bennett, Wendell C. *Ancient Arts of the Andes*. New York: The Museum of Modern Art, 1954.

_____ . "Weaving in the Land of the Incas." *Natural History* 36 (1935), 63–72.

_____ and Bird, Junius B. *Andean Culture History*. 2d ed. New York: American Museum of Natural History, 1960.

Bhushan, Jamila Brij. *The Costumes and Textiles of India*. Bombay: D. B. Taraporevala Sons & Co., 1958.

Bird, Junius. "America's Oldest Farmers." *Natural History* 57 (1948), 296–303ff.

_____ . "Before Heddles Were Invented." *Handweaver & Craftsman* 3 (Summer 1952), 5–7ff.

_____ and Mahler, Joy. "America's Oldest Cotton Fabrics." *American Fabrics*, No. 20 (Winter 1951–52), 73–79.

Birrell, Verla. *The Textile Arts : A Handbook of Fabric Structure and Design Processes*. New York: Harper and Bros., 1959.

Bishop, J. Leander. *A History of American Manufactures from 1608 to 1860*. Vol. 1. Philadelphia: Edward Young, 1861.

Blumenau, Lili. *The Art and Craft of Handweaving*. New York: Crown, 1955.

Broholm, H. C., and Hald, Margrethe. *Costumes of the Bronze Age in Denmark; contributions to the archaeology and textile-history of the bronze age*. Translated by Elizabeth Aagesen. Copenhagen: Arnold Busck, 1940.

Bühler, Alfred. "The Development of Weaving Among Primitive Peoples." *Ciba Review*, No. 30 (1940), 1058–85.

Bühler, Kristin. "Basic Textile Techniques." *Ciba Review* No. 63 (1948), 2290–2320.

Burnham, Harold B. "Çatal Hüyük—The Textiles and Twined Fabrics." *Anatolian Studies* 15 (1965), 169–74.

_____ . "Four Looms." University of Toronto, The Royal Ontario Museum, Art and Archaeology Division, *Annual* (1962), 77–84.

_____ and Burnham, Dorothy K. *'Keep Me Warm One Night': Early Weaving in Eastern Canada*. Toronto: University Press, 1972.

Carroll, Diane Lee. "The Heddle in Greek Art." Paper presented at 67th General Meeting of the Archaeological Institute of America (Dec. 28–30, 1965).

Carter, H. R. "Spinning and Weaving in Early Times." *Scientific American, Supplement* 82 (1916), 154–55.

Carus-Wilson, Eleanora. "The Woolen Industry," *The Cambridge Economic History of Europe*. Vol. 2 (Trade and Industry in the Middle Ages). Cambridge: The University Press, 1952.

Cherblanc, Émile. "Mémoire sur l'invention du tissu." *Histoire générale du tissu*, Document I. Paris: Les Éditions d'Art et d'Histoire, 1935.

Christensen, Bodil. "Otomi Looms and *Quechquemitls* from San Pablito, State of Pueblo, and from Santa Ana Hueytlapan, State of Hidalgo, Mexico." *Carnegie Institution of Washington, Division of Historical Research, Notes on Middle American Archaeology and Ethnology*, No. 78 (1947), 122–42.

Clark, J. G. D. *Prehistoric Europe: The Economic Basis*. Stanford: Stanford University Press, 1966.

Clark, Robert Judson, ed. *The Arts and Crafts Movement in America, 1876–1916*. Princeton: Princeton University Press, 1972.

Collingwood, Peter. "Neolithic Weaving Techniques." *Quarterly Journal of the Guilds of Weavers, Spinners and Dyers*, No. 35 (1960), 174–79.

Cordry, Donald Bush and Cordry, Dorothy M. "Costumes and Weaving of the Zoque Indians of Chiapas, Mexico." *Southwest Museum Papers*, No. 15 (1941).

_____ . *Mexican Indian Costumes*. Austin: University of Texas Press, 1968.

Crane, Walter. "On the Revival of Design and Handicraft." *Arts and Crafts Essays by Members of the Arts and Crafts Exhibition Society*. London: Longmans Green & Co., 1903.

Crawford, M. D. C. *The Conquest of Culture*. New York: Greenberg, 1938.

_____ . *5000 Years of Fibers and Fabrics*. Exhibition handbook. Brooklyn, N.Y.: Brooklyn Museum, Brooklyn Institute of Arts and Sciences, 1946.

_____ . *The Heritage of Cotton: The Fiber of Two Worlds and Many Ages*. New York: Fairchild, 1948.

_____ . "The Loom in the New World." *The American Museum of Natural History Journal* 16 (1916), 381–87.

_____ . "Peruvian Textiles." *American Museum of Natural History, Anthropological Papers*, Vol. 12, Pt. 3 (1915).

Crowfoot, Grace M. "Spinning and Weaving in the Sudan." *Sudan Notes and Records* 4 (1921), 20–38.

_____ . "Textiles, Basketry, and Mats." *A History of Technology*. Vol. 1. Edited by Charles Singer, E. J. Holmyard, and A. R. Hall. Oxford: The Clarendon Press, 1954.

_____ . "The Vertical Loom in Palestine and Syria." *Palestine Exploration Quarterly* (Oct. 1941), 141–51.

_____ . "Of the Warp-Weighted Loom." *The Annual of the British School at Athens*, No. 37 (1936–37), 36–47.

Cyrus, Ulla. *Manual of Swedish Hand Weaving*. Boston: Charles T. Branford, 1958.

Diderot, Denis. *A Diderot Pictorial Encyclopedia of Trades and Industry: Manufacturing and the Technical Arts in Plates Selected from "L'Encyclopédie, ou Dictionnaire Raisonné des Sciences, des Arts et des Metiers."* 2 vols. Edited by Charles Coulston Gillispie. New York: Dover, 1959.

Douglas, Frederic H. "Indian Cloth-Making: Looms, Technics, and Kinds of Fabrics." *Denver Art Museum, Dept. of Indian Art, Leaflets* 59–60, Denver, Colo. (1933).

Drum, Jim. "Andean Weaving Draws on the Past." *El Palacio*, Vol. 81, No. 4 (1975), 35–45.

Earle, Alice Morse. *Home Life in Colonial Days*. New York: Macmillan, 1898.

Eco, Umberto and Zorzoli, G. B. *The Picture History of Inventions*. New York: Macmillan, 1963.

Emery, Irene. *The Primary Structures of Fabrics*. Washington: The Textile Museum, 1966.

Emmons, George T. "The Chilkat Blanket." *American Museum of Natural History, Memoirs*, Vol. 3, Pt. 4 (1907), 329–50.

Faxon, Harriet. "A Model of an Ancient Greek Loom." *Metropolitan Museum of Art, Bulletin*, Vol. 27, No. 3, Sec. 1 (1932), 70–71.

Flanagan, J. F. "Figured Fabrics." *A History of Technology*. Vol. 3. Edited by Charles Singer, E. J. Holmyard, and A. R. Hall. Oxford: The Clarendon Press, 1957.

_____ . "The Origin of the Drawloom Used in the Making of Early Byzantine Silks." *The Burlington Magazine* 35 (1919), 167–72.

Forbes, Robert J. *Studies in Ancient Technology*. Vol. 4. Leiden: E. J. Brill, 1956.

Geijer, Agnes. "The Loom Representation on the Chiusi Vase." *Studies in Textile History*. Edited by Veronika Gervers. Toronto: Royal Ontario Museum, 1977.

_____ and Anderbjörk, Jan Erik. "Two Textile Implements from the Early Middle Ages." *Folk-Liv* 3 (1939), 232–41.

Gilroy, Clinton G. *The Art of Weaving by Hand and by Power*. New York: George D. Baldwin, 1844.

Glazier, Richard. *Historic Textile Fabrics*. London: B. T. Batsford, 1923.

Goris, Dr. R. *Bali: Atlas Kebudajaan: Cults and Customs*. Government of the Republic of Indonesia, 1956.

Gutmann, A. L. "Cloth-Making in Flanders." *Ciba Review*, No. 14 (1938), 466–90.

_____ . Varron, A., and Schaefer, Ch. G. "Tapestry." *Ciba Review*, No. 5 (1938), 138–71.

Hald, Margrethe. "Olddanske Tekstiler." *Nordiske Fortidsminder*, Vol. 5 (1950).

_____ . "Olddanske Tekstiler: fund fra aarene 1947–55." *Aarbøger for Nordisk Oldkyndighed og Historie*, (1955), 1–60.

Hamblin, Dora Jane. *The First Cities*. New York: Time-Life Books, 1973.

Harcourt, Raoul d'. "Peruvian Textile Techniques." *Ciba Review*, No. 136 (1960), 2–40.

_____ . *Textiles of Ancient Peru and Their Techniques*. Edited by Grace G. Denny and Carolyn M. Osborne. Translated by Sadie Brown. Seattle: University of Washington Press, 1962.

Henshall, Audrey S. "Textiles and Weaving Appliances in Prehistoric Britain." *Proceedings of the Prehistoric Society*, N. S. 16 (1950), 130–62.

Hindson, Alice. *Designer's Drawloom*. Boston: Charles T. Branford, 1958.

Hoffmann, Marta. *The Warp-Weighted Loom*. Oslo: Universitetsforlaget, 1964.

_____ and Burnham, Harold. "Prehistory of Textiles in the Old World." *Viking* 37 (1973), 49–76.

Holmes, W. H. "Prehistoric Textile Art of Eastern United States." *Smithsonian Institution, Bureau of American Ethnology, 13th Annual Report* (1896), 9–46.

Hooper, Luther. *Hand-Loom Weaving: Plain and Ornamental*. London: Sir Isaac Pitman & Sons, 1920.

_____. "The Loom and Spindle: Past, Present, and Future." *Smithsonian Institution, Annual Report of the Board of Regents for 1913-14* (1915), 629-78.

_____. "The Technique of Greek and Roman Weaving." *Burlington Magazine* 18 (1911), 276-84.

Innes, R. A. *Non-European Looms*. Halifax, Eng.: Halifax Museums, 1959.

Irwin, John. "Indian Textiles in Historical Perspective." *Marg* 15 (1962).

Iyer, D. S. V. "Looms." *Marg* 15 (1962).

Jaques, Renate and Flemming, Ernst. *Encyclopedia of Textiles: Decorative Fabrics from Antiquity to the Beginning of the 19th Century Including the Far East and Peru.* New York: Praeger, 1958.

James, George Wharton. *Indian Blankets and Their Makers*. New York: Tudor Publishing Co., 1937.

Jayakar, Mme. Pupul and Irwin, John. *Textiles and Ornaments of India*. Edited by Monroe Wheeler. New York: The Museum of Modern Art, 1956.

Jones, Volney H. "A Summary of Data on Aboriginal Cotton of the Southwest." *University of New Mexico Bulletin*, No. 296, *Anthropological Series* 1 (1936), 51-64.

Joyce, Thomas A. "Babunda Weaving." *Ipek: Jahrbuch für Prähistorische & Ethnographische Kunst*, (1925), 105-10.

_____. "The Peruvian Loom in the Proto-Chimu Period." *Man* 21 (1921), Art. 106, 177-80.

Kahlenberg, Mary Hunt and Berlant, Anthony. *The Navajo Blanket*. New York: Praeger Publishers with Los Angeles County Museum of Art, 1972.

Keller, Dr. Ferdinand. *The Lake Dwellings of Switzerland and Other Parts of Europe*. 2 vols. London: Longmans, Green & Co., 1878.

Kent, Kate Peck. "A Comparsion of Prehistoric and Modern Pueblo Weaving." *The Kiva*, Vol. 10, No. 2 (1945).

_____. "The Cultivation and Weaving of Cotton in the Prehistoric Southwestern United States." *American Philosophical Society, Transactions*, n. s. 47, Pt. 3 (1957).

_____. *Introducing West African Cloth*. Denver: Denver Museum of Natural History, 1971.

King, Mary Elizabeth. "The Prehistoric Textile Industry of Mesoamerica." Paper presented at Textile Museum Roundtable, Washington, D. C. (1975).

Kissell, Mary Lois. *Yarn and Cloth Making*. New York: Macmillan, 1918.

Klein, O. "Textile Arts of the Araucanians." *Ciba Review*, 1961/6 (1961), 2-25.

Lamb, Venice. *West African Weaving*. London: Duckworth, 1975.

_____ and Alastair Lamb. *The Lamb Collection of West African Narrow Strip Weaving*. Edited by Patricia Fiske, Washington: The Textile Museum, 1975.

Leggett, William F. *The Story of Linen*. Brooklyn, N. Y.: Chemical Publishing Co., 1945.

_____. *The Story of Silk*. New York: Lifetime Editions, 1949.

Leix, Alfred. "Ancient Egypt, the Land of Linen." *Ciba Review*, No. 12 (1938), 397-404.

Lewis, Ethel. *The Romance of Textiles*. New York: Macmillan, 1938.

Little, Frances. *Early American Textiles*. New York: The Century Co., 1931.

Liu, Gaines K. C. "The Silkworm and Chinese Culture." *Osiris* 10 (1952), 129-93.

Lopez, Robert S. *The Commercial Revolution of the Middle Ages, 950-1350*. Englewood Cliffs, N.J.: Prentice-Hall, 1971.

Lucas, A. *Ancient Egyptian Materials and Industries*. 3d ed. London: Edward Arnold & Co., 1948.

Mace, A. C. "Loom Weights in Egypt." *Ancient Egypt,* Pt. 3 (1922), 75-76.

Mason, J. Alden. *The Ancient Civilizations of Peru*. Baltimore, Md.: Penguin, 1968.

Mason, Otis T. "A Primitive Frame for Weaving Narrow Fabrics." *Smithsonian Institution, U. S. National Museum, Annual Report for 1899* (1901), 487-510.

_____. *Woman's Share in Primitive Culture*. New York: D. Appleton & Co., 1894.

Matthews, Washington. "Navajo Weavers." *Smithsonian Institution, Bureau of Ethnology, 3rd Annual Report* (1884), 375-91.

Means, Philip Ainsworth. *Ancient Civilizations of the Andes*. New York: Charles Scribner's Sons, 1931.

Naylor, Gillian. *The Bauhaus*. London: Studio Vista, 1968.

Needham, Joseph. *Science and Civilization in China*, Vol. 4, Pt. 2 (Mechanical Engineering). Cambridge: The University Press, 1965.

Nettleship, Martin A. "A Unique South-East Asian Loom." *Man, Journal of the Royal Anthropological Institute* V (1970), 686-698.

Nylen, Anna-Maja. *Swedish Handcraft*. New York: Van Nostrand Reinhold Co., 1977.

O'Neale, Lila M. "Mochica (Early Chimu) and Other Peruvian Early Twill Fabrics." *Southwestern Journal of Anthropology* 2 (1946), 269-94.

_____. "Textiles of Highland Guatemala." *Carnegie Institution of Washington, Pub.* 567 (1945).

_____. "Weaving." "Handbook of South American Indians." Vol. 5. Edited By J. H. Steward. *Smithsonian Institution, Bureau of American Ethnology,* Bull. 143 (1949), 97-138.

Olson, Ronald L. "The Possible Middle American Origin of Northwest Coast Weaving." *American Anthropologist*, n. s. 31 (1929), 114-21.

Patterson, R. "Spinning and Weaving." *A History of Technology*. Vol. 2. Edited by Charles Singer, E. J. Holmyard, and A. R. Hall. Oxford: The Clarendon Press, 1956.

_____. "Spinning and Weaving." *A History of Technology*. Vol. 3. Edited by Charles Singer, E. J. Holmyard, and A. R. Hall. Oxford: The Clarendon Press, 1957.

Plumer, Cheryl. *African Textiles: An Outline of Handcrafted Sub-Saharan Fabrics*. Michigan: Michigan State University, 1971.

Polkinhorne, R. K. *Weaving and Other Pleasant Occupations*. New York: Brentano, 1923.

Postan, Michael. "The Trade of Medieval Europe: the North." *The Cambridge Economic History of Europe*. Vol. 2 (Trade and Industry in the Middle Ages). Cambridge: The University Press, 1952.

Reichard, Gladys A. *Navajo Shepherd and Weaver*. New York: J. J. Augustin, 1936.

Richter, Gisela M. A. "A Stand by Kleitias and an Athenian Jug." *Metropolitan Museum of Art, Bulletin* (1931), 289–94.

Riesenberg, Saul H. and Gayton, A. H. "Caroline Island Belt Weaving." *Southwestern Journal of Anthropology* VIII (1952), 342–375.

Roth, H. Ling. "Ancient Egyptian and Greek Looms." 2d ed. *Bankfield Museum Notes*, 2d. ser., No. 2 (1951).

_____ . *Studies in Primitive Looms. Bankfield Museum Notes*, 2d ser. Nos. 3–11 (1918). Reprint, New York: Burt Franklin Reprints, 1974.

Roth, Walter Edmund. "Additional Studies of the Arts, Crafts, and Customs of the Guiana Indians." *Smithsonian Institution, Bureau of American Ethnology*, Bull. 91 (1929).

Salzman, L. F. *English Industries of the Middle Ages.* Oxford: The Clarendon Press, 1923.

Schaefer, Gustav. "The Loom." *Ciba Review,* No. 16, (1938), 542–77.

Schuette, Marie. "Tablet Weaving." *Ciba Review,* No. 117 (1956), 2–29.

Sieber, Roy. *African Textiles and Decorative Arts.* New York: Museum of Modern Art, 1972.

Siewertsz van Reesema, Elizabeth. "Contribution to the Early History of Textile Technics." *Verhandelingen der Koninklijke Akademie van Wetenschappen*, n.s. Vol. 26, No. 2 (1926).

Spier, Leslie. "Zuni Weaving Technique." *American Anthropologist,* n. s. 26 (1924), 64–85.

Start, Laura E. "Indian Textiles from Guatemala and Mexico." *Man* 48, Art. 78, (1948), 67–68.

Sylwan, Vivi. "Investigation of Silk from Edsen-gol and Lop-nor." *Reports from the Scientific Expedition to the North-Western Provinces of China under the Leadership of Dr. Sven Hedin, The Sino-Swedish Expedition, Pub.* No. 32, 7. Archaeology: 6 (1949).

_____ . "Silk from the Yin Dynasty." *Museum of Far Eastern Antiquities, Bulletin*, No. 9 (1937), 119–26.

Thomson, W. G. *A History of Tapestry: From the Earliest Times Until the Present Day.* 2 vols. London: Hodder and Stoughton, 1930.

Tryon, Rolla Milton. *Household Manufactures in the United States, 1640–1860.* Chicago: University of Chicago Press, 1917.

Underhill, Ruth. *The Navajos.* Norman, Okla.: University of Oklahoma Press, 1956.

_____ . "Pueblo Crafts." *Indian Handcrafts* 7. Education Division, U. S. Indian Service, Washington, D. C. [1945].

Usher, Abbott Payson. *A History of Mechanical Inventions.* Rev. ed. Cambridge: Harvard University Press, 1954.

Varron, A. "The Early History of Silk." *Ciba Review*, No. 11 (1938), 350–88.

Vogt. E. "Basketry and Woven Fabrics of the European Stone and Bronze Ages." *Ciba Review,* No. 54 (1947), 1938–72.

Vollmer, John E. "Archaeological and Ethnological Considerations of the Foot-Braced Body-Tension Loom." *Studies in Textile History.* Edited by Veronika Gervers. Toronto: Royal Ontario Museum, 1977.

Walbank, Frank William. "Trade and Industry under the Later Roman Empire in the West." *The Cambridge Economic History of Europe.* Vol. 2 (Trade and Industry in the Middle Ages). Cambridge: The University Press, 1952.

Walton, Perry. *The Story of Textiles.* New York: Tudor Publishing Co., 1936.

Warden, Alexander J. *The Linen Trade, Ancient and Modern.* London: Longman, Green, Longman, Roberts & Green, 1864.

Watson, J. Forbes. *The Textile Manufactures and the Costumes of the People of India.* London: George Edward Eyre and William Spottiswoode, 1866.

Weibel, Adele Coulin. *Two Thousand Years of Textiles.* New York: Pantheon Books, 1952.

Weir, Shelagh. *Spinning and Weaving in Palestine.* London: The British Museum, 1970.

Wescher, H. "The Cloth Trade and the Fairs of Champagne." *Ciba Reivew*, No. 65 (1948), 2362–96.

Westheim, P. "Textile Art in Ancient Mexico." *Ciba Review,* No. 70, (1948), 2554–86.

Wild, J. P. *Textile Manufacture in the Northern Roman Provinces.* Cambridge: The University Press, 1970.

Williams, M.C. "The Homespun Age." *Magazine of American History with Notes and Queries* 25 (1891), 239–43.

Wilson, Lillian M. *The Clothing of the Ancient Romans.* The Johns Hopkins University Studies in Archaeology, No. 24. Baltimore: The Johns Hopkins Press, 1938.

_____ . "Loom Weights." *Excavations at Olynthus.* Pt. 2. Edited by David M. Robinson. The Johns Hopkins University Studies in Archaeology, No. 9. Baltimore: The Johns Hopkins Press, 1930.

Winlock, H. E. "Heddle-Jacks of Middle Kingdom Looms. *Ancient Egypt,* Pt. 3 (1922), 71–74.

_____ . *Models of Daily Life in Ancient Egypt.* Cambridge: Harvard University Press and Metropolitan Museum of Art, 1955.

Wulff, Hans E. *The Traditional Crafts of Persia.* Cambridge: The M. I. T. Press, 1966.

Yates, James. *Textrinum Antiquorum: An Account of the Art of Weaving Among the Ancients.* London: Taylor and Walton, 1843.

Index